The Cultural Value of Trees

This volume focuses on the tree, as a cultural and biological form, and examines the concept of folk value and its implications for biocultural conservation.

Folk value refers to the value of the more-than-human living world to cultural cohesion and survival, as opposed to individual well-being. This field of value, comprising cosmological, aesthetic, eco-erotic, sentimental, mnemonic value and much more, serves as powerful motivation for the local performance of environmental care. The motivation to maintain and conserve ecology for the purpose of cultural survival will be the central focus of this book, as the conditions of the Anthropocene urgently require the identification, understanding and support of enduring, self-perpetuating biocultural associations. The geographical scope is broad, with chapters discussing different tree species from the Americas and the Caribbean, East Asia, Eurasia and Australia and Africa. By focusing on the tree, one of the most reliably cross-culturally-valued and cross-culturally-recognized biological forms, and one which invariably defines expansive landscapes, this work illuminates how folk value binds the survival of more-than-human life forms with the survival of specific peoples in the era of biocultural loss, the Anthropocene. As such, this collection of cross-cultural cases of tree folk value represents a low hanging fruit for the larger project of exploring the power of the cultural value of the more-than-human living world.

This book will be of great interest to students and scholars of conservation, biodiversity, biocultural studies and environmental anthropology.

Jeffrey Wall is the managing editor for the Journal of Ethnobiology and holds a PhD in natural resources and the environment from Cornell University, USA.

The Earthscan Forest Library
Series Editorial Advisers:
John L. Innes
University of British Columbia, Canada
John Parrotta
US Forest Service – Research & Development, USA
Jeffrey Sayer
University of British Columbia, Canada
and
Carol J. Pierce Colfer
Center for International Forestry Research, USA

This series brings together a wide collection of volumes addressing diverse aspects of forests and forestry and draws on a range of disciplinary perspectives. It is aimed at undergraduate and postgraduate students, researchers, professionals, policy-makers and concerned members of civil society.

Forest Conservation and Sustainability in Indonesia
A Political Economy Study of International Governance Failure
Maxton Bernice-Lee

Masculinities in Forests
Representations of Diversity
Carol J. Pierce Colfer

Adaptive Collaborative Management in Forest Landscapes
Villagers, Bureaucrats and Civil Society
Edited by Carol J. Pierce Colfer, Ravi Prabhu, and Anne M. Larson

Sacred Forests of Asia
Spiritual Ecology and the Politics of Nature Conservation
Edited by Chris Coggins and Bixia Chen

The Cultural Value of Trees
Folk Value and Biocultural Conservation
Edited by Jeffrey Wall

For more information about this series, please visit: www.routledge.com/books/series/ECTEFL

The Cultural Value of Trees

Folk Value and Biocultural Conservation

Edited by Jeffrey Wall

Routledge
Taylor & Francis Group
LONDON AND NEW YORK

earthscan
from Routledge

First published 2022
by Routledge
4 Park Square, Milton Park, Abingdon, Oxon OX14 4RN

and by Routledge
605 Third Avenue, New York, NY 10158

Routledge is an imprint of the Taylor & Francis Group, an informa business

British Library Cataloguing-in-Publication Data
A catalogue record for this book is available from the British Library

Library of Congress Cataloging-in-Publication Data
A catalog record for this book has been requested

ISBN: 978-0-367-33615-8 (hbk)
ISBN: 978-1-032-26519-3 (pbk)
ISBN: 978-0-429-32089-7 (ebk)

DOI: 10.4324/9780429320897

Typeset in Bembo
by Apex CoVantage, LLC

To Asildar, my dear friend and teacher. May blessings be upon your family. Long live the Avar. May Car thrive.

Contents

Figures and tables

Figures

Tables

Foreword

Markers of the first mass seeing – Albert Namatjira observing country and gum trees

I write this foreword to *The Cultural Value of Trees: Folk Value and Biocultural Conservation* as a non-Aboriginal, from settler culture in Australia. I have been privileged to have worked alongside the descendants of the famous Australian watercolour artist Albert Namatjira co-creating theatre works and documentaries and other initiatives with the family and, at times, writing and speaking on their behalf, with their permission. This foreword is a reflection on my time working with the family and the culture, the many experiences of learning and exchange which occurred, as well as his remarkable life's work painting the landscapes of central Australia, including his trademark ghostly white river gums.

The privilege of working with his descendants came as a gift through my work of thirty years with the non-profit organisation, Big hART. Big hART's purpose is to reveal urgent and often invisible stories which need to be told, so as to help restore cultural justice. At the centre of this work is the motivating phrase, "it is harder to hurt someone if you know their story." Big hART makes content to help build community and drive change. These urgent and invisible stories are found in many diverse places, and over the years, invitations to co-create have come from many First Nations people – Western Aranda, Pitjantjatjara, Aranda, Ngarluma, Yindjibarndi, Yuin, Wiradjiri and Gadigal

Albert Namatjira holds an iconic place in contemporary Australia's cultural history, and this centrality lent an urgency to the telling of the almost unknown story of his family's loss of ownership of the copyright of his pictures, to assist in restitution and cultural justice. Working with the community at Ntaria in Central Australia, Big hART commenced an eight-year project comprised of community workshops, touring exhibitions, a media strategy, a sell-out theatre show, short films, a lobbying strategy, an audience with the Queen and Prince Phillip and a documentary, so as to bring this important story to public attention and drive change. The first step in this co-creation journey was to sit down as a team with the Namatjira family to paint with watercolours. It is very difficult. Strangely, to try and paint a ghost gum in its landscape requires an act of stillness, and of trying to enter some knowledge of country. It requires almost not painting, leaving the white of the parchment to represent the white bark on the trunk of the tree. It requires I think, a new kind of observance.

Albert Namatjira was born in 1902, around the time when the six states of Australia were federated into one nation, and in this way, Albert's (Elea – Western Aranda name) life traced the trajectory of modern Australia. He walked out of the bush as a two-year-old, to camp near a German Lutheran mission, and as grown-up he straddled two divergent cultural realities: one in which he was considered a powerful man of Western Aranda culture, and the other in which he was infantilised, othered, punished, paraded and treated by the law as though he was part of the flora and fauna.

In this confused context, Albert came across a remarkable new technology, a new media (watercolour) which allowed him to represent his surroundings in literal ways and preserve it, creating magic to the eyes of his countrymen, who, for the first time, saw magical manifestations of the colour blue. They saw new representations of their country, not the deep and private worlds of ceremony and mark making, something other, derived of another purpose – to reproduce the surface as it was and hold it in your hands. To carry it with you. To place it on a wall, as if to create a portal into the outside world. It is hard to describe the impact of this vibrant technology. It was, however, seen as a cultural undertaking, and therefore linked into the whole of life on country, where art and country and family and law are not naively separated and made into commodities as they are in the West.

Albert easily combined these ancient and contemporary purposes, painting totemic trees on country while holding back representations of culturally dangerous and sacred places, and produced remarkable watercolour works from his country, for acquisition by these excitable and curious non-Aboriginal people who, for the most, part hugged the safety of the coast, thousands of miles from Albert's country. These works were relatively quickly discovered by these city folk, and he was soon held in awe nationally – both loved and patronised, collected and trashed. In the process, he painted his family out of extreme poverty, at the height of his fame, supporting 600 family members with his brush and his eye, and feeding Australia's thirst for his subtle, eerie landscapes, which somehow we knew were beckoning us to belong.

At the heart of his remarkable body of work, the 'Namatjira trees' stand like sentinel beings, framing the landscape of these portals. To the non-Aboriginal eye, they remain to this day deeply iconic as well as politically, culturally and socially charged. It is almost as if we have always seen them, when in fact, Albert made us see them, these tall, thick-trunked beings, reaching deep underground and revelling in the harsh conditions of the desert. These painted trees were, in their time, arresting invitations for each wave of migrants, making our way here, across oceans to a new world and new hope, and as prints of Namatjira's landscapes appeared behind counters on the walls of government interiors for quarantine, immigration, education and services, they almost became invitations welcoming and signifying arrival and destiny, like windows into the soul of this alien new landscape.

Albert's life, beginning in the bush with almost no contact with white people, became the life of an icon, a rock star, and yet ended with wrongful imprisonment, a spent spirit, death soon after and then the near destruction of his legacy through poor critique and the loss of the copyright for his family. Yet those trees and their watercolour representation have lived on. His totemic images crossing cultural, societal, geographic boundaries, contributing to the social myth-making, from the coastal rind of continent to Namatjira country in the centre. The myth-making has been done by the citizenry, songwriters and artists of this modern Australia, who have rarely seen the 'outback' of Central Australia – miss-named as Namatjira's country – and serves a hunger for belonging. Perhaps clumsily, it serves to define what it means to be Australian in the midst of our colonisation and unrelenting oppression, persisting despite the growing desire for recognition and acknowledgement of traditional owners.

And so, when we place our feet on country and feel it as we read this, when we observe our country, when we allow Namatjira's totemic trees – and all the other generous Indigenous intercultural and hybrid art making – to speak to us, it becomes clear that we are being welcomed into country, with seemingly perpetual generosity and forgiveness, and that observance and acknowledgement of country is much more than a kind of cultural good manners. Observance of country is about the survival of the planet and humanity.

We are not observing country through a lens of remorse – sensing death, despair, deficit, sorrow, defeat, inferior literacies or the ease of a dying pillow. We are turning the lens on ourselves and sensing our own deficits. In Namatjira's trees, we are sensing the knowledge, belonging, joy, continuity, stewardship, diplomacy, sustainability, food, organics, precision gathering and farming, harmony, bio-diversity, leisure, sensuality, eroticism, feasting and the everyday festival of Indigenous life. These are two very different versions of the same narrative – one leads to tacit slow genocide, problematising and offering solutions, the other leads to continuity and sustainability.

In the face of the timelessness Albert's landscapes capture, we are so fleeting. Jung put it like this, "Life is a luminous pause between two great mysteries that in the end are one" (Jung 1947). There is something about this luminosity, that is related to country, to belonging, to our place in the biosphere, to our contribution as big-brained primates – there is a deep resonance between Jung's phrasing, and the frequent aboriginal phrasing "being on country" a kind of observance during our moment of luminosity. And this luminosity is what sets Namatjira's pictures apart. It is my pleasure to write the opening words and thoughts to *The Cultural Value of Trees*, which begins the work of acknowledging the priceless luminosity offered by the lives of trees and the vitality of our worlds' Indigenous Peoples to the destiny of our planet and our species.

Scott Rankin

Note

At the conclusion of Big hART's eight-year project working with the Namatjira family, the copyright has been returned and a trust set up for royalties and donations to assist in well-being, education and literacy. We observe deeply everything we learned from collaborating on this project.

References

Jung, Carl. 1947. *Letters*, vol. 1, edited by Gerhard Adler, Aniela Jaffé, translated by R. F. C. Hull. Princeton University Press, Princeton, NJ.

Contributors

Scott Rankin, Director, Big hART, Australia

Jeffrey Wall, Postdoctoral Associate, Conservation through Reconciliation Partnership, University of Guelph, Department of Geography, Environment and Geomatics & Visiting Assistant Professor of Environmental Studies, Hobart and William Smith Colleges

José Tomás Ibarra, Department of Ecosystems and Environment, School of Agriculture and Forestry Sciences & Center of Applied Ecology and Sustainability (CAPES), Pontificia Universidad Católica de Chile, Santiago, Chile & ECOS (Ecosystem – Complexity – Society) Co-Laboratory, Center for Local Development (CEDEL) & Center for Intercultural and Indigenous Research (CIIR), Villarrica Campus, Pontificia Universidad Católica de Chile, Villarrica, Chile & Cape Horn International Center for Global Change Studies and Biocultural Conservation (CHIC).

Robert Petitpas, Department of Geography, University College London, London, UK

Antonia Barreau, ECOS Co-Laboratory, CEDEL & CIIR, Villarrica Campus, Pontificia Universidad Católica de Chile, Villarrica, Chile

Julián Caviedes, ECOS Co-Laboratory, CEDEL & CIIR, Villarrica Campus, Pontificia Universidad Católica de Chile, Villarrica, Chile & Institut de Ciència i Tecnologia Ambientals, Universitat Autònoma de Barcelona, Barcelona, Spain

Josefina Cortés, ECOS Co-Laboratory, CEDEL & CIIR, Villarrica Campus, Pontificia Universidad Católica de Chile, Villarrica, Chile

Gabriel Orrego, ECOS Co-Laboratory, CEDEL & CIIR, Villarrica Campus, Pontificia Universidad Católica de Chile, Villarrica, Chile

Gonzalo Salazar, CEDEL & CIIR, Villarrica Campus, Pontificia Universidad Católica de Chile, Villarrica, Chile

Tomás A. Altamirano, ECOS Co-Laboratory, CEDEL & CIIR, Villarrica Campus, Pontificia Universidad Católica de Chile, Villarrica, Chile

Patricia Shanley, People and Plants International, Woods & Wayside International, Ridgeview Conservancy, Princeton, New Jersey, USA

Frederica Bowcutt, Member of the of Faculty of Botany, The Evergreen State College, Washington, USA

John Rashford, Professor Emeritus, Department of Sociology and Anthropology, College of Charleston, South Carolina, USA

Marie J. Zahn, School of Aquatic and Fishery Sciences, University of Washington, USA

Joe Martin, Tla-o-qui-aht First Nation

Eli Enns, IISAAK OLAM Foundation and the Indigenous Protected and Conserved Areas (IPCA) Innovation Program

Peter S. Wyse Jackson, President, Missouri Botanical Garden, St Louis, MO, USA, and George Engelmann Professor of Botany, Washington University in St Louis

Amots Dafni, Professor Emeritus, University of Haifa, Israel

Saleh ‘Aql Khatib, Mghar, Israel

Fabio Gatti, International Institute of Social Studies (The Hague), Erasmus University Rotterdam, The Netherlands

Asaad Habibi, Department of Horticulture, College of Aburaihan, University of Tehran, Iran

Mohsen Hamedpour-Darabi, Department of English Language and Literature, Azarbaijan Shahid Madani University, Tabriz, Iran

Kourosh Vahdati, Department of Horticulture, College of Aburaihan, University of Tehran, Iran

Roy Ellen, Centre for Biocultural Diversity, School of Anthropology and Conservation, University of Kent, Canterbury, Kent, UK

Andrea Flores Urushima, Kyoto Seika University, Kyoto, Japan

Benoit Jacquet, École française d'Extrême-Orient (EFEO), Paris, France

Wil de Jong, Kyoto University, Japan

Romain Simenel, l'Institut de Recherche pour le Développement (IRD) I, Patrimoines locaux, Environnement et Globalisation (PALOC), UMR 208, France

Yildiz Aumeeruddy-Thomas, French National Centre for Scientific ResearchCentre (CNRS), D'ecologie Fonctionnelle & Evolutive (CEFE), UMR 5175, France

Ahmed Aarab, University Abdelmaleck Es-saadi, Tanger, Morocco

Morgan L. Ruelle, Clark University Department of International Development, Community and Environment, Massachusetts, USA

Zemede Asfaw, College of Natural Sciences, Addis Abbaba University, Ethiopia

1 Tree folk value

Seeing a thriving living world through cultural survival

Jeffrey Wall

> The belief of all, faith, is the effect of the need of all, of their unanimous desires.
> – Marcel Mauss and Henri Hubert (Mauss and Hubert 1904)

One desire of this volume is to be all about the force held within Albert Namatjira's painted visions of ghostly gums. That is, the paradigm-blasting fire of an Australian Aboriginal man's fervent remembrance of trees through paints. As our foreword author recounts, the fact that these paintings will not be reproduced here in accordance with law is testimony to that cultural fire, to the heat of that fire and the fuel load it consumed. We must imagine his trees because we shall not acquire them. These bright gums, like Albert's people and kin, are of their land and vision and shall remain so. Unlike the prancing citizens of Kosztka's Edenic fever dream (see Figure 1.1), the Western Aranda shout back and cause our social imagination to open.

Another desire is to emerge from this shared meditation on the strong bond between people and trees with solemn, pragmatic insights for a humanity bursting into its own geological era like a meteorite, purging earth's vitality and its own nature in its throes. As a committed biocultural text, the interests of humanity's more muted constituents – namely, Indigenous Peoples, cultural minorities and local communities – take precedence. The environmental goods we are in a position to steward are a legacy of their lifeways. Their cultural survival is essential to continued environmental health at the level of plot and planet. For it is a settled matter that these "heterogenous, disparate" (Graeber 2001: 56) peoples harbor the heterogenous and disparate values needed to maintain meaningful life from the white noise of earth-alienated, universalized, value. Stated another way, this volume joins a growing conviction that the "icy waters of egotistical calculation," (Marx and Engels 1848: 16) even should they blast away in the name of biodiversity protection, will protect very little without the power of more widely-held desires. In short, this volume is organized and prepared to deliver robust testimony to the power of cultural survival to power the environmental care needed in our new era.

Folk value is the term rendered to refer to this value of the more-than-human living world to cultural cohesion and survival, as opposed to individual

DOI: 10.4324/9780429320897-1

Figure 1.1 Pilgrimage to the Cedars of Lebanon, Csontváry Kosztka (1907).

well-being. This field of value serves as powerful motivation for the local performance of biological conservation. The motivation to maintain and conserve ecology for the purpose of cultural survival will be the central focus of this book, as the conditions of the Anthropocene urgently require the identification, understanding and support of enduring, self-perpetuating biocultural associations. By focusing on the tree, one of the most reliably cross-culturally-valued and cross-culturally-recognized biological forms, and one which defines expansive landscapes and cultures alike, this work aims to illuminate how folk value binds the survival of more-than-human life forms to the survival of specific peoples in the face of persistent biocultural loss during the Anthropocene.

The concept of folk value is a timely and important innovation in sustainability studies for the following reasons. First, in the face of painful evidence that economic value cannot be harnessed for the purposes of stemming environmental degradation and species loss, a larger exploration and analysis of value which can rival the pre-eminence of market value is necessary. A new, cultural, framework for value has the potential to operationalize deep, self-perpetuating value for environmental well-being. Second, the plain linkage of the Anthropocene and the sixth mass extinction reveals that biological objects – such as specimens, species, ecosystems and landscapes which are *not* bolstered by anthropogenic forces –

face accelerating risk of disappearance. Due to their iconic, multi-functional, aesthetic and symbolic natures, trees are unparalleled among bio-forms around the world for the ubiquity, intensity and charisma to peoples of the world. As such, this cross-cultural, multi-case exploration of tree folk value around the world represents low hanging fruit for the larger project of supporting the strong and vital bonds stowed in cultural value for the more-than-human living world.

Our immanent frame

Ours is an era of immeasurable biological loss caused by the human species. The dawning, twin, realities of the Anthropocene (Crutzen 2006) and the sixth mass extinction (Barnosky et al. 2011), taken together, reveal that those parts of the biosphere not bolstered by anthropogenic forces face accelerating risk of extinction. In this context, identifying and supporting enduring and exquisite biocultural associations – or meaningful patterns of interaction, exploitation and maintenance between a people and a biological community – represents a premier strategy for conserving biological richness and complexity.

However, there are urgent and critical constraints on this approach. For one, cultural diversity is also in steep decline, and this decline, through a mist of under-studied – and probably unstudy-able – mechanisms, is inextricably related to the loss of biological diversity (Loh and Harmon 2014). Likewise, the provision of cultural erosion is not random; it is the cultures most intensively associated with their local ecologies, namely Indigenous and Tribal Peoples, who are most likely to face cultural extinction (Fleming and Ledogar 2008). These factors, and many more, demonstrate the necessity of achieving simultaneous biological and cultural conservation which meets certain standards of ethics and efficacy. Yet the languages and models of biological conservation and cultural conservation are uniquely incompatible. To illustrate, asking what it means for a culture to survive and for an individual to do the same conjures entirely distinct modes of thought. In the first case, an individual is thought to be surviving, in conventional wisdom, if their heart is beating, though they may be comatose. In other words, individual survival is commonly known as a medical if not mechanical diagnosis. A culture, on the other hand, is thought to be surviving if some unknowable recipe of its institutional, linguistic, culinary and aesthetic traditions are being maintained by a number of individuals deemed sufficient. The designation of cultural survival, rightfully or not, lends itself to more interdisciplinary modes of thought. This book deals squarely with this dissonance by considering the trees that humankind needs to survive meaningfully rather than metabolically.

Deeper understandings of value for biocultural conservation

An anthropological approach to value differs from economics chiefly due to its inherent endeavor to engage, understand and interpret the perceptions and contexts of those whose culture is different from the community of inquirers.

For the purposes of this book, it logically follows that culturally valued trees have been shaped over millennia in a dynamic with the values of their associated human communities. What is important about an anthropology of value is its illumination of value heterogeneity in the ethnographic record. In the broadest sense that it exists today, anthropology of value lands on the study of three phenomena: 1) the role of sociality in circulation of value, 2) cultural performance by which value comes to exist in media and 3) processes of value redemption between different media (Graeber 2001). Insights from these three pursuits, fleshed out with ethnographic evidence, render an anthropology of value a formidable foil to assumptions that *homo economicus* existed in pre-industrial society, and tailored his plants and landscapes with the respective premium on productivity.

Branislow Malinowski, Marcel Mauss and Karl Marx represent the pantheon of such an understanding of value. Malinowski is the famous marooned anthropologist, whose visa quandary during World War II caused him to reside for over two years with the Trobriand Islanders of today's Papua New Guinea. During this time, Malinowski had ample time to observe, ponder and document the elaborate and mystifying exchange of *Kula*, decorative strings of shells, which were exchanged between households and communities, but rarely if ever worn. What struck Malinowski, and what is an essential data point for an anthropology of value, is that this exchange system provided a rival model of currency to that held in modern economies. Trobriand Islanders, with Kula, as well as with other resources such as livestock and garden space, put a premium on painstaking aesthetic efforts which stanched productivity and efficient accumulation. The engine of Kula exchange, according to Malinowski, was not material accumulation, it was the prestige accrued through impressive, elaborate public giving in combination with heroic and risk-intensive ocean travel. (Malinowski 1922)

Mauss fixated on the transubstantive phenomena by which value came to inhabit media. One of his most famous bodies of work on this subject, *The Gift*, dealt with the strong undercurrents within gift exchange which bid the recipient of a gift to reciprocate. His approach to understanding this power was to review the available ethnographic accounts of three cultural cases, the Kula Ring on the Trobriand Islands, the Potlach of the Northwestern North American Indigenous Kwakiutl and the gift giving traditions of the Maori people of Aotearea. His central assertion was that gift objects are everywhere perceived/known to absorb and carry some piece or amount of their owner. Therefore, a gift, once given, exerts a desire to return to its owner while simultaneously embodying the interests and sentience of the giver as long as it remains in the possession of the recipient. He further posits, much like Malinowski, that those hoping to identify the germ of *homo economicus* in pre-industrial exchange will come away empty handed. An enduring contribution of Mauss regarding value was showing that the sterile and conclusive exchange model central to utilitarian economics, if it is to be found at all in the ethnographic record, can be traced to the class of trade relations between enemies, or between those to whom the other's fate is of no importance (Mauss 1954; Graeber 2001).

Marx, in his ponderous studies of industrial capitalist economies, provides a third leg to an anthropological theory of value. His theoretical approach was founded on his labor theory of value; namely that value could be known as the proportion of a society's labor devoted to the manufacture or maintenance of a good, service or other entity. Marxian anthropologists have carved out a niche for themselves by studying such principles at work in 'pre-industrial' societies. United by an evaluation of societies according to production and labor, this research is responsible for a noteworthy insight. Across much of the world, in societies most peripheral to globalized markets, the highest proportion of labor is devoted, not to commonly understood 'economic' endeavors like food production and item manufacturing, but is instead devoted to the manufacture of satisfactory persons. In other words, child-rearing, facilitation of rites of passage, celebrations and the like commonly occupy much more social time and labor (Turner 1979; Munn 1986; Fajans 1997). According to Marx, and to common sense as well, the vast majority of labor in capitalist societies is devoted to the manufacture of commodities and to the delivery of services. Thus, while the production of human persons is no doubt necessary in this system, their value is redeemed in the form of commodities and services. Marxian ethnography has uncovered that the vast majority of pre-industrial societies demonstrate the precise opposite pattern; namely, labor which has been devoted to the 'economic' is redeemed in the form of socially-determined quality persons. This would naturally include labor spent modifying and working with surrounding life forms like trees.

The ethnobotanical record offers countless observations of values other than utilitarian instrumentalist ones that define human-plant relations. The following three consensuses found in the ethnographic record will prove useful in illustrating the significant diversity in values across societies. The first is that management and selection of specific varieties often take place for religious and cosmological purposes. For example, in West Java, Somarwoto documents how landraces are linked to unique individual and household identities based on perceived similarities in what might be described as personality traits (Soemarwoto 2007). Here, selection of rice varieties for planting is decided through consideration of the variety's 'affective' (p. 91) qualities, followed by the physiological and culinary. Affective traits include thoughts, ideas and personal styles attributed to the rice variety as well as the variety's eligibility as the object of certain rituals performed in various life stages of the plant and crop. Similarly, Quechua farmers in the Andes cultivate the *culli* landrace of maize specifically as a household protector that is hung on the doorframe of a kitchen (Hilgert et al. 2013). Furthermore, in early colonial Tenochtitlan, it was observed that an incredible premium was put on the earliest ripening varieties of amaranth for the manufacture of idols cannibalized ritually in the harvest festivities (Early 1992).

A second consensus is distinct but intersects with the first and last. The botanical holdings of diaspora groups, which originate in peoples' native lands, are understood to represent a very valuable selection of flora for the translocated groups. With the example of African populations in the Americas, plant holdings which originate in Africa are often found in the plant-use repertoires of Maroon, freed slave and other African-descended communities (Voeks and Rashford 2012). Such plant species, which were trafficked despite tremendous hardship and risk, commonly have a multiplicity of properties including nutritional, ecological and spiritual. Examples include two native African tree species that are now well established in the new world, the African oil palm (*Elaeis guineensis*) and the baobab tree (*Adansonia digitate*). At the dawn of the slave trade, Indigenous coastal West Africans, especially those groups around the Gulf of Guinea, held the African oil palm as an invaluable species for nutrition and spirituality. Similar circumstances evolved in the Brazilian region of Bahia, where the locally prominent Afro-Caribbean religion of Candomblé features a number of specific uses for materials from the tree (Watkins 2011, 2015). The baobab tree has been observed in a number of African diaspora geographies in the Americas. In addition to a number of edible parts, including fruits, seeds and flowers, the tree's physiology manifests a completely distinctive ecology which houses bees, attracts bats, collects and holds water in its trunk and, in otherwise bare landscapes, provides copious shade. The tree is integral in numerous folkloric and spiritual traditions in both Africa and the new world (Rashford 1987a, 1987b, 2015).

The obvious and unique value for the Baobab may also lend support to a third consensus from the ethnobotanical record and that is that aesthetic value for plants, including crops, was and is a pervasive factor in their natural history. The baobab's appearance, distinct and touching by any account, makes it a revered icon of the African savannah landscape (Rashford 2015). This undoubtedly played a role in its urgent selection for transport to the new world. Aesthetic properties have been integral to crop selection and evolution throughout history (Hawkes 1983; McCouch 2004). Aesthetics remain the most significant historical driver of exotic plant trade in the modern era (Mack and Lonsdale 2001). Returning to amaranth, Aztec royalty demanded beauty from the crop to such an extent that the transfixed Spanish transported amaranth to Spain, initially as an ornamental, apparently unaware that the crop was one of four grain crops acquired as tribute from Aztec territory (Early 1992). This pattern of guaranteeing beauty through plant selection has been observed with crops throughout the Americas (Debouck 1989). Stepping back from the ethnobotanic record, it is a matter of common sense that the physical appearance of plants, including food plants, is of considerable day-to-day importance.

Key insights from this volume

Material presented in this volume adds to the crest of this cross-cultural testimony to value for more-than-human life. The collective value refractions which the previous analysis highlights, cosmological and aesthetic, are more fully substantiated through the studies contained in this volume. Furthermore, the shared focus of our project has put into relief more areas worth considering, namely mnemonic, eco-erotic value and the value of plants like trees to social solidarity and resistance.

Aesthetic value, or sensorial delight, takes a particularly interesting turn in Ruelle and Zemede's overview of the non-native, cultivated "new king" eucalyptus, which dominates the sensorial milieu of highland Debark, Ethiopia to a degree that is hard to imagine. However, favor and delight over fragrance, thatching, siding, ground texture and more lag far behind raw utility at the present moment. The utility of this this tree reaches such heights that it is said to be priming local people for their inevitable reverence, that is, the "new king." Similarly, Ellen shows how the low aesthetic ring to the sago palm for the Polynesian Nuaulu of Polynesia gives way to the intimate familiarity and dependability of this species, leaving it the core means of environmental knowledge and social ties. Similar evolutionary momentum is documented in my own chapter on the chestnut in Turkey, where aesthetics are observed to be a prelude to and stimulation toward a more ultimate value of shared enchantment. Compounding aesthetic provocations like these in each context, instead of making straight lines, amounts to something like waves, which manifest as Ellen eloquently puts it, in "interlocking dependencies" (p. 181).

Numerous instances of inscription or mental-biological hybridity occur in these pages as well. As Simenel, Aumeeruddy-Thomas and Aarab lay out, healthy, beloved and manicured argan groves are themselves known to be texts holding the history of former matrilineal lifeways, djinn mythologies and kinship records for all who care to see. Likewise, Urushima, Jaacquet and de Jong illustrate how the socio silvicultural fabric of today's Kitayama Sugi forestry provides a clear and detailed testimony of the Edo period, especially visible in its contrast with today's aging and sparse rural populations and the widespread dwindling of traditional tea culture. Wyse Jackson meanwhile traces a unique thread of Irish cultural evolution through the social life of the hawthorn, from a medieval tree alphabet known as Oghan, through Irish resistance against the British and into today's environmentalist blogosphere. These accounts, along with others, reveal the striking range of cultural energies stowed in our landscapes and surrounding life forms.

One of the more unique messages in this text relates to the connective tissue between humanity's encounter with life forms and with the divine. Both Dafni and Khatib as well as Rashford categorically reject any notion of tree worship in their studies, seeing trees in a more invocational role. Nyame's altar, as Rashford dubs the cotton tree in Jamaica, presents a channel to divinity and in some instances also to the famed cultural resilience of the Ashanti peoples of Ghana. Likewise, through a dogged review of religious texts in combination with field interviews, Dafni and Khatib show that while the variability of positive versus negative spiritual associations for the carob tree vary over time and space, at the same time, there remains a striking consistency to the intensity of attitudes. Finally, Zahn, Martin and Enns beat a path away from a crisis narrative for the decline of the red cedar near the Nuu-Chah-Nulth community along the west coast of Vancouver Island. Instead, they draw on Indigenous philosophy, which itself originates in red cedar forest, to envision respect for the "firm but loving hand" of climate change, which lies behind the dwindling prospects of the red cedar. These considerations, among others, bid readers to vigilantly consult trees and their devoted communities for enduring visions of strength, courage and love.

Additionally, many authors read tales of solidarity and resistance into the tree-people associations they studied. Bowcutt shows the pepperwood tree at the beating heart of California Indian resistance to colonialism, bio-prospecting, real estate speculation and boutique food trends. Agents she encounters across history and into the present day dress themselves in the bioprosperity of pepperwood in their struggle to survive and keep culture. Trees rallying their people to resistance is rather a constant in this book. The hawthorn fairy trees famously draw protection of locals against urban sprawl; God trees in Jamaica serve as venues for conspiracy by night; eucalyptus in Debark stakes out sovereignty of farmers in the face of incoherent forestry policy. Last but certainly not least, the white pine – the former symbol of enduring peace for the Haudenosaunee Confederacy – shows up ominously at the ransacking of the US Capitol on January 6, 2021. Like with Namatjira, the firepower of trees lights up the page again and again.

Finally, one of the fortunate opportunities for this volume is the chance to work with what Melissa Nelson dubs 'eco-erotics' (2017). While her original formulation revels in erotic tales, visions and experiences between people and more-than-human nature, for the purposes of this book it was useful to locate eroticism in Carl Jung's notion of libido (1967). Jung attempted to gain latitude beyond Freud on this particular idea. He did this by arguing that libido, though undeniably linked to biological instincts around procreation, was above all a much larger and holistically motivating, outer-directed, "life-urge" (1967: 136). This urge promised to fulfill a need to engage the world beyond the individual and perhaps the society. Thus, "the world is empty only to him who does not know how to direct his libido toward things and people and to render them alive and beautiful" (1967: 173). This notion that engagement with and intentional impact of human persons on their world is unstoppable and inherently good, has implications far and wide.

It is a certainty that much erotic content in the ethnosciences has been suppressed. This volume catches glimpses in Habibi, Hamdpour-Darabi and Vahdati's reporting of a gonad association for the walnut in Iran, and in the aphrodisiac potency of Hawthorn blossoms outside as Wyse Jackson documents. Additionally, my own work in Turkey took place for years before I was informed that the word for chesnut, *kestane*, was euphemistic for vagina due to the visual resemblance conjured by the opened and spiny husk. However, in the broad libidinal sense offered by Jung, trees' receptivity to shepherding, manipulation, positioning and more makes this eco-erotic value a constant through these pages. As Ibarra et al. eloquently describe, this mixture amounts to "sympoietic relations" – habits of becoming together (p. 24 of this volume). This *sympoiesis* lends itself to a very special insight into the charisma of trees throughout human history and into our Anthropocene, namely that their synanthropic leanings have never gone unnoticed. Theirs is a sharing and inviting world we ought to bear forward.

Folk value and the great curation

The Anthropocene and its conjoined twin, the Sixth Mass Extinction of species, once acknowledged, freight in other, more tectonic, acknowledgments. First, they require an acknowledgment that social processes and human attributes preserve the biosphere to the extent that it is preserved. In this understanding, trees, forests, wilderness, biological diversity, ecosystem sustainability and other objects of conservation exist where, and to the extent that, human value is expressed.

The second implicit acknowledgment is that unconscious and conscious selection comingle to perform the winnowing process. The Norwegian wharf rat, the starling and kudzu do not owe their prevalence to conscious selection. On the other hand, peculiar examples of conscious selection abound. Crowns-of-thorns starfish-killing robots patrol the Great Barrier Reef and Barred owls are shot by rangers in the Northwest United States to provide relief for the spotted owl. In the vast majority of cases, human-caused extinctions are most easily understood as unconscious acts. The elimination of numerous species caused by the decreased area of the Amazon Forest is, itself, an repercussion unintended effect. Yet, how unconscious would the act of harvesting the horn of the last northern white rhinoceros be? Clearly, extinctions in our era occur according to hybrid conscious and unconscious forces.

The third implicit acknowledgment is that value will determine the proportionality of unconscious to conscious acts of species deletion. Value exists at distinct but interacting levels. From a meta perspective, in our era, the valuation inherent to the globalized economy is the most powerful organizing principle. Carbon emissions, circulation of invasive pathogens, ecological fragmentation: such biologically deleterious phenomenon, often referred to as externalities, are generated by countless acts occurring in the pursuit of value. Pursuit of value at the global level relies on the routine establishment and maintenance of certain species and populations, while simultaneously eliciting ambivalence toward and deletion of others. Within sovereign nations, values may be expressed in the face of globalized market dynamics which may or may not contrast with market value. Threats to certain known rivers, certain species and certain tree populations are noted or not noted, responded to or not responded to with available means. The world over, heterogeneous fields of value that hover below the national level are routinely muted.

The final implicit acknowledgment is that such value is not being lost by everyone equally. There is hierarchy inherent to the Anthropocene and to this great curation. The truth of this is borne out by a particular and commonly cited ethical observation. The cultural groups least responsible for global climate change, that is, Indigenous and Traditional Peoples, presently experience some of its earliest, most direct and brutal effects. Extinction and other forms of biological erosion are integral parts of this injustice. This book puts forward a concrete approach for conscious and moral deciphering of loss within this great curation through a concerted focus on one treasured, and threatened, bioform across the world: trees. This collection of cases shows that there is voluminous undocumented care and value for the biosphere powering the daily lives of people and peoples across the world. Similar efforts to understand and give voice to this rooted, diverse and durable value are imperative for at least two reasons. First, it may amplify and disseminate long-muted, but highly motivating, value for the biosphere. Second, these heterogenous and disparate values provide essential guidance for seeing and safeguarding the optimal biological value in our great curation.

References

Barnosky, A. D., N. Matzke, S. Tomiya, G. O. U. Wogan, B. Swartz, T. B. Quental, C. Marshall, J. L. McGuire, E. L. Lindsey, K. C. Maguire, B. Mersey, and E. A. Ferrer. 2011. Has the Earth's Sixth Mass Extinction Already Arrived? *Nature* 471: 51–57. https://doi.org/10.1038/nature09678.

Crutzen, P. 2006. The "Anthropocene". In *Earth System Science in the Anthropocene*, pp. 13–18. Springer, Berlin Heidelberg.

Debouck, D. 1989. Early Beans (Phaseolus vulgaris L. and P. lunatus L.) Domesticated for their Aesthetic Value. In *Annual Reptort of Bean Improvement Cooperative*. Bean Improvement Cooperative, Mayaguez, Puerto Rico, USA.

Early, D. 1992. The Renaissance of Amaranth. In *Chilies to Chocolate: Food the Americas Gave the World*, edited by N. Foster and L. Cordell. University of Arizona Press, Tucson, AZ.

Fajans, J. 1997. *They Make Themselves: Work and Play among the Baining of Papua New Guinea*. University of Chicago Press, Chicago, IL.

Fleming, J., and R. J. Ledogar. 2008. Resilience, an Evolving Concept: A Review of Literature Relevant to Aboriginal Research. *Pimatisiwin* 6: 7–23.

Graeber, D. 2001. *Toward an Anthropological Theory of Value: The False Coin of Our Own Dreams*. Palgrave Macmillan, New York, NY.

Hawkes, J. G. 1983. *The Diversity of Crop Plants*. Harvard University Press, Cambridge, MA.

Hilgert, N. I., F. Zamudio, V. Furlan, and L. Cariola. 2013. The Key Role of Cultural Preservation in Maize Diversity Conservation in the Argentine Yungas. *Evidence-Based Complementary and Aternative Medicine* 2013: 1–10. https://doi.org/10.1155/2013/732760.

Jung, Carl. 1967. *Symbols of Transformation: An Analysis of the Prelude to a Case of Schizophrenia*. Princeton University Press, Princeton, NJ.

Kosztka, Csontváry. 1907. Pilgrimage to the Cedars of Lebanon. *Google Arts & Culture*.

Loh, J., and D. Harmon. 2014. *Biocultural Diversity: Threatened Species, Endangered Languages*. WWF, Netherlands.

Mack, R. N., and W. M. Lonsdale. 2001. Humans as Global Plant Dispersers: Getting More than We Bargained For. *BioScience* 51: 95. https://doi.org/10.1641/0006-3568(2001)051[0095:HAGPDG]2.0.CO;2.

Malinowski, B. 1922. *Argonauts of the Western Pacific: An Account of Native Enterprise and Adventure in the Archipelagoes of Melanesian New Guinea*. Routledge, London.

Marx, K., and F. Engels. 1848. Manifesto of the Communist Party. *Marx Today: Selected Works and Recent Debates*: 1–263. https://doi.org/10.1057/9780230117457.

Mauss, M. 1954. *The Gift*. Cohen and West, New York, NY.
Mauss, M., and H. Hubert. 1904. Outline of a General Theory of Magic. *L'Annee Sociologique* 7: 1–146.
McCouch, S. 2004. Diversifying Selection in Plant Breeding. *PLoS Biology* 2: 1507–1512. https://doi.org/10.1371/journal.pbio.0020347.
Munn, N. 1986. *The Fame of Gawa: A Symbolic Study of Value Transformation in a Massim (Papua New Guinea) Society*. Cambridge University Press, Cambridge, UK.
Nelson, M. 2017. Getting Dirty: The Eco-Eroticism of Women in Indigenous Oral Literatures. In *Critically Sovereign: Indigenous Gender, Sexuality, and Feminist Studies*, edited by J. Barker, pp. 229–260. Duke University Press, Durham, NC.
Rashford, J. 1987a. The Search for Africa's Baobab Tree in Jamaica. *Jamaica Journal* 20: 2–11.
Rashford, J. 1987b. The Baobab Tree and Seasonal Hunger Africa: The Case of the San. *Botswana Society* 19: 57–68.
Rashford, J. 2015. The Uses of the Baobab Flower (Adansonia digitata L.). *Ethnobotany Research and Applications* 14: 211–229. https://doi.org/10.17348/era.14.0.211-229.
Soemarwoto, R. 2007. Kasepuhan Rice Landrace Diversity, Risk Management and Agricultural Modernization. In *Modern Crises and Traditional Strategies: Local Ecological Knowledge in Island Southeast Asia*, edited by R. Ellen, pp. 84–111. Berghahn Books, New York.
Turner, T. 1979. Kinship, Household and Community Structure among the Kayapo. In *Dialectical Societies: Ge and Bororo of Central Brazil*, edited by D. Maybury-Lewis, pp. 179–217. Harvard University Press, Cambridge, MA.
Voeks, R., and J. Rashford. 2012. *African Ethnobotany in the Americas*, edited by R. Voeks and J. Rashford. Springer Science and Business Media, New York, New York.
Watkins, C. 2011. African Oil Palm Agroecologies in Bahia, Brazil, and Implications for Development. *Journal of Latin American Geography* 10: 9–33. https://doi.org/10.1353/lag.2011.0005.
Watkins, C. 2015. African Oil Palms, Colonial Socioecological Transformation and the Making of an Afro-Brazilian Landscape in Bahia, Brazil. *Environment and History* 21: 13–42. https://doi.org/10.3197/096734015X14183179969700.

Part I

The Americas and the Caribbean

2 Becoming tree, becoming memory

Social-ecological fabrics in Pewen (*Araucaria araucana*) landscapes of the southern Andes

José Tomás Ibarra, Robert Petitpas, Antonia Barreau, Julián Caviedes, Josefina Cortés, Gabriel Orrego, Gonzalo Salazar and Tomás A. Altamirano

Introduction

Trees are rooted in a myriad of biocultural landscapes worldwide (Watkins 2016; Holmgren and Scheffer 2017). They play key ecological roles such as the creation of microhabitats for biodiversity, carbon sequestration, and the maintenance of hydrological, nutrient, and energy cycles (Lindenmayer and Laurance 2017). Trees have also long been symbols within complex bodies of human knowledge, practices, and beliefs, as they provide tangible goods (resin, bark, flowers, wood, fruits, medicines, and seeds) and intangible values (aesthetic, artistic, and religious character) (Hageneder 2009; Kohn 2013). These roles are continually unfolding in situated social-ecological systems, in which properties of the system emerge from localized interactions that contribute to the specific fabrics of differing landscapes' milieus (Filotas et al. 2014; Ibarra et al. 2020b).

These principles speak through the poet Leonel Lienlaf (1989) when he states:

> *La vida del árbol invadió mi vida. Comencé a sentirme árbol*
> *y entendí su tristeza.*
> *Empecé a llorar por mis hojas, mis raíces,*
> *mientras un ave*
> *se dormía en mis ramas esperando que el viento dispersara sus alas.*
> *Yo me sentía árbol porque el árbol era mi vida.*
>
> The life of the tree invaded my life.
> I started to feel like a tree and I understood its sadness.
> I started crying through my leaves, my roots,
> while a bird
> was falling asleep in my branches waiting for the wind to scatter its wings. I became a tree because the tree was my life.
> (Lienlaf 1989)

DOI: 10.4324/9780429320897-3

An outstanding property of social-ecological fabrics is the biocultural memory of the system, which emerges from the stored experiences and information (e.g., evolutionary, ecological, historical) shared by living beings with one another and with their territories (Toledo and Barrera-Bassols 2008; Ibarra et al. 2021). These records of historical events continue to influence the compositional, structural, and functional states of the system (Ibarra et al. 2020b). For example, in an ecological context, seeds are repositories of memory as they are full embryonic plants responsible for the nourishment of the embryo, the dispersal to a new location, and for becoming dormant during unfavorable environmental conditions (Nazarea 2005). For their part, large ancient trees retain the memory of past forest conditions in their tree rings and canopy structure, and often sustain complex biodiversity networks in forested landscapes (Briffa 2000; Ibarra et al. 2020b). In a social context, trees and their seeds are intertwined with many people's livelihoods and intergenerational processes of knowledge acquisition and senses of place, thus giving and negotiating meanings of human experience with other humans and other-than-human beings (Jones and Cloke 2002; Kohn 2013).

In current studies on social-ecological systems linked with trees, we commonly find two approaches. The first focuses on biological entities, examining the ecological dynamics of tree species and associated biodiversity, and the combination of structures that make reorganization possible (Ibarra et al. 2020b). This approach is generally built on variable-driven analyses, searching for cause-effect relationships among components that allow the functioning of the system (de Vos et al. 2019); ecological dynamics are generally conceived to be shaped or disrupted by anthropogenic drivers of change (e.g., climate change, deforestation, habitat degradation, fires, etc.) (Caviedes and Ibarra 2017). A second approach focuses on people, analyzing human agency along with the historical and contemporary political drivers of change (e.g., power relations) shaping human-tree landscapes (Jones and Cloke 2002). This approach commonly explores people's perceptions and attitudes, and how their knowledge, practices, and beliefs influence their historical and present relations with trees (Watkins 2016). A third perspective, named the relational approach, highlights integrated social-ecological relations to overcome the ecological/social and the biological entity/human agency dichotomies of the two previous approaches (Murdoch 1997; Darnhofer et al. 2016). As relations are continually reshaped through ecological and social dynamics, the biocultural memory of the system emerges from the ability to make and remake relations (Ibarra et al. 2021).

The Pewen (*Araucaria araucana*) represents one of the last remnants of the ancient, and formerly widespread, coniferous family Araucariaceae in South America (Figure 2.1). This species, considered to be a true 'living fossil,' remains in a global biodiversity hotspot centered in the temperate rainforests of Chile and Argentina (Donoso 1993). The Pewen reaches up to 30 meters in height and can age more than 1,500 years. It tends to form pure canopies above an intermediate stratum of broadleaf trees (*Nothofagus* spp.) in forests shaped by humans for millennia (Maestre et al. 2012). Indeed, there has been a permanent presence

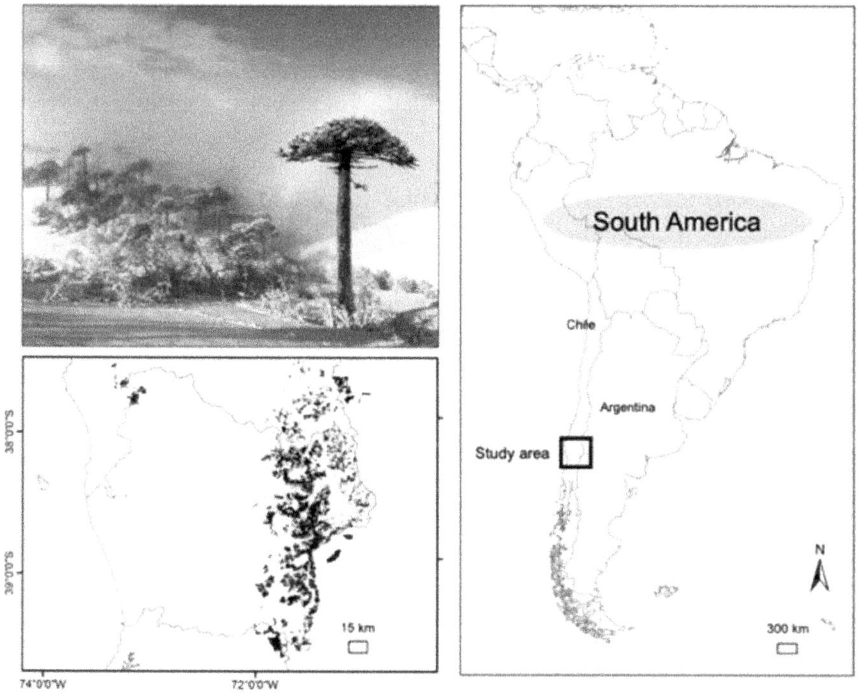

Figure 2.1 Study area in the La Araucanía Region (bottom left), showing the current extent (shaded black) of Pewen (*Araucaria araucana*) forests in the southern Andes of South America.

of human beings in the distribution range of Pewen landscapes since 4,000 BP (Reis et al. 2014). Indigenous groups facilitated the expansion of *Araucaria* forests across South American landscapes during the Holocene. But, after European colonization in Chile and Argentina in late 19th century, and mostly during the 20th century, fires from these new settlers and the logging industry cleared nearly 50% of Pewen forests (González et al. 2005; Armesto et al. 2010). Remaining Pewen populations, classified globally as endangered, are threatened by climate change, fire, overgrazing, exotic seed predators, and lack of recruitment (IUCN 2021). The importance of the Pewen would not only lie in its botanical and ecological value (Ibarra et al. 2010; Cockle et al. 2019), but also in its cultural (religious, economic, educational, commercial) importance for the Mapuche-Pewenche Indigenous people (*Pewen*: Araucaria and *Che*: people; Pewenche = People of the Pewen) (Herrmann 2005, 2006; Barreau et al. 2016, Cortés et al. 2019). The Pewenche are the Mapuche who inhabit the Andes mountain range in south-central Chile, where they co-inhabit with Pewen forested landscapes.

In this chapter, we examine social-ecological systems associated with the Pewen by critically reflecting and integrating the results of our long-term

(2008–2021) inter- and transdisciplinary work in the southern Andes of Chile. We first briefly describe our studies on biological entities and their interaction fabrics in Pewen forests, examining the ecological dynamics of trees and associated biodiversity. Then, we recount our work on Mapuche-Pewenche social fabrics linked to the Pewen which are constrained, but not necessarily determined, by historical and contemporary drivers of change. Finally, we elaborate a relational approach, which highlights unfolding and integrated social-ecological relations, overcoming both the ecological/social and the biological entity/human agency dichotomies. We discuss this relational approach by examining the biocultural memory of the system, which is not a fixed asset but a property of the system that emerges from situated social-ecological relations that are continuously made and remade with both past and current circumstances.

Setting the landscape: forests and people

Since 2008, we have studied Pewen social-ecological fabrics in the Andean zone of the La Araucanía region of Chile, which is part of the *Wallmapu*, the territory of the Mapuche people (Figure 2.1). At lower elevations, the landscape is characterized by valleys floors with a heterogeneous mosaic of human settlements, farmlands, fragments of native forest, agroforestry systems, plantations of non-native trees, shrublands, lakes, and rivers. As the elevation rises, native forest increasingly dominates the landscape. At these higher elevations, Pewen emerges progressively mixed with evergreen Coihue beech (*Nothofagus dombeyi*) at elevations 900–1,100 meters above sea level (masl) and with deciduous Lenga beech (*Nothofagus pumilio*) and Antarctic beech (*Nothofagus antarctica*) at 1,100–1,600 masl. Mapuche lands and communities are interspersed with farms owned by non-Mapuche settlers. These farms are often larger than the communities, as the Mapuche have seen their access to land disrupted by private properties and a history of State colonization policies (Huiliñir-Curío 2010), including the creation of a number of protected areas (Barreau et al. 2016). As a result of decades of discrimination, the local language, *Mapuzugun*, was progressively lost, being now mostly spoken by elders. Younger members of Mapuche communities speak mostly in Spanish and basic *Mapuzugun*. However, there is rising awareness and numerous efforts and initiatives to recover the language.

Throughout the years, we have deployed several biodiversity survey and ethnographic methodologies. For ecological research, we have used bird point counts, mist-netting, camera trapping, peeper cavity-inspecting cameras, reptile counts through transects, and vegetation plots, among others (e.g., Ibarra et al. 2010, 2017; Ibarra and Martin 2015; Altamirano et al. 2017; Caviedes and Ibarra 2017; Cockle et al. 2019). For social inquiring, we have implemented, for example, participant observation, oral histories, interviews, free-listing, food elicitation diaries, focus groups, and other means of participatory action research (e.g., Barreau et al. 2016, 2019; Cortés et al. 2019; Ibarra et al. 2021). All these methodologies have allowed us to get a broad and deep understanding of, as well as a long-term commitment to, social and ecological fabrics in Pewen landscapes.

Biological entities: Pewen ecological fabrics

Prototypical biological entities in forested landscapes are the community of wildlife species that use tree-cavities to achieve their life history requirements. Cavity-using wildlife interact in fabrics known as 'nestwebs' in the ecological literature (Martin and Eadie 1999). These fabrics are generally viewed as hierarchical and commensal, with resource (cavities) flowing from trees to non-excavators (e.g., many songbirds, ducks, raptors, and other organisms), facilitated by the action of excavators (e.g., woodpeckers) and tree-decay processes (Figure 2.2; Martin et al. 2004).

In high-elevation Andean forests (>900 masl), we have reported that Pewen is one of the dominant trees, as it accounts for 30% of forest basal area. However, it provides only 2% of nesting cavities (Cockle et al. 2019). There are 26 bird species, six mammals, and at least two reptile species that breed in tree cavities in Pewen forests (Figure 2.2; Altamirano et al. 2017). Nests are placed in all tree species present in these forests, but most secondary cavity-nesting species use

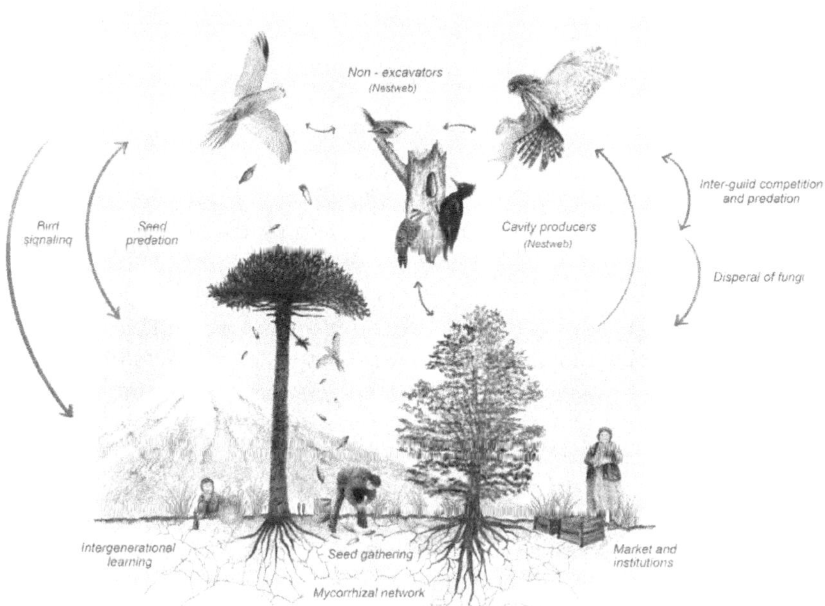

Figure 2.2 Pewen landscapes in the southern Andes integrate multiple entities/agents that co-produce each other in complex and sympoietic social-ecological fabrics that vary across space (from the seed to the country and beyond) and time (building on and revisiting the past). Thus, the Pewen, as it is currently, is but a stable moment in a continuing process of 'becoming' (see main text for details on the arrows and specific social-ecological interactions).

Source: Illustration by M. Medina.

cavities formed by tree-decay processes in *Nothofagus* spp. (98% of nest cavities in Pewen forests). Broadleaf trees that coexist with Pewen, especially the Lenga beech, thus play a key role at the base of complex nestwebs of interacting birds, mammals, reptiles, and trees. The Austral parakeet (*Enicognathus ferrugineus*) is the only species showing a relatively important use of cavities produced by Pewen (11% of its nests; Díaz and Kitzberger 2013). Austral parakeets also play an important role in facilitating seed germination and regeneration of Pewen (Speziale et al. 2018). Therefore, the Austral parakeet, through its feeding and nesting habits, is an example of a biological entity that links Pewen and *Nothofagus* trees as they feed on and disperse Pewen seeds and, simultaneously, nest chiefly in cavities available in coexisting old-growth *Nothofagus* trees in the southern Andes. The latter example highlights that cavity nesters interact through mutualisms, facilitation, competition, and predation, as well as through the dispersal of seeds of future nest trees (Blanc and Walters 2007; Ibarra et al. 2014; Altamirano et al. 2017). Furthermore, cavity nesters disperse spores of wood-decay fungi that facilitate cavity-formation as well (Jusino et al. 2016), while some species may also reduce recruitment of cavity trees through seed predation (Figure 2.2; Shepherd et al. 2008).

In Pewen landscapes, repositories of memory in ecological fabrics include, for example, long-lived system entities such ancient Lenga beech, Coihue beech, Pewen, and forest patches whose presence and influence extend over a long time. Indeed, in these old-growth forest stands, the cavity-tree species with the highest density of useful cavities for wildlife are large decaying trees of Lenga beech (66% of all cavities) and Coihue beech (16%); species that accompany Pewen trees (Ibarra et al. 2020c). Furthermore, these fabrics are influenced by disturbance events such as fires, with wildlife community responses depending heavily on time since the fire. These responses are related to the progress of vegetation succession with an associated gradual increase in structural complexity of Pewen landscapes (Infante et al. 2021; Novoa et al. 2021).

The above-ground community of cavity nesters is connected to a below-ground fabric of mycorrhizal fungi that links current and future generations of trees (Simard et al. 2013). Mycorrhiza is the symbiotic association between plant roots and fungi, and it is present in most plants around the world. Plants can translocate up to 30% of the carbohydrates to their root-associated fungi. In return, fungi behave as skilled root extensions that explore and interact formidably with soil, nurturing and protecting their host. Furthermore, the exploring mycelium of the fungi can associate with several other trees simultaneously, enabling forest trees to physically link together and establish complex below-ground communication networks (Molina and Horton 2015; Simard 2018). In Pewen trees, studies have shown a high degree of mycorrhizal fungal colonization on their roots (Diehl and Fontenla 2010; Godoy and Marín 2019). Mycorrhiza is particularly important to Pewen trees after fires, because the symbiotic fungi extend their range of exploration, finding and uptaking scarce and immobilized nutrients such as phosphorus (Figure 2.2; Diehl and Fontenla 2010).

Despite that mycorrhizal networks in Pewen forests have not been explored in detail, the formation of a common mycelium that links trees within and between species has been reported for temperate forests elsewhere (Smith and Read 2010). Simard et al. (1997) showed evidence of bidirectional inter-plant carbon transfer between conifer and deciduous broadleaf trees. This may be a dynamic occurring between Pewen and coexisting *Nothofagus* species. These links, nevertheless, might occur differently in southern temperate forests because Pewen and *Nothofagus* associate with different groups of mycorrhizal fungi; Pewen is arbuscular and *Nothofagus* is ectomycorrhizal. There is no evidence of arbuscular and ectomycorrhizal mycelium linking together and establishing common pathways between plants. However, it has been suggested that *Nothofagus* trees can be dual mycorrhizal, potentially associating with arbuscular and ectomycorrhizal mycelium (Smith and Read 2010; Teste et al. 2020). While mycorrhizal fabrics still need to be examined in Pewen ecosystems, it might be hypothesized that *Nothofagus* trees may facilitate the access of Pewen to a larger pool of nutrients. Ectomycorrhizal fungi have higher capabilities to capture organic nutrient and are able to form more extensive mycelium across the forest ground compared to arbuscular fungi (Talbot et al. 2008; Carteron et al. 2020). Nevertheless, arbuscular fungi persist or recover better than ectomycorrhizal fungi after fire events. Therefore, the empirically validated and hypothesized interconnectedness of the biological entities (fungi, trees, cavity nesters) and dynamics in Pewen forests underlie their nature as self-organized complex adaptive systems (Simard et al. 2013; Ibarra et al. 2020b). Here, processes as disparate as disturbances, facilitation, competition, dispersal, and nutrient cycling are connected through cross-scale below- and above-ground fabrics.

Human agency: Pewen social fabrics

> because before, everyone ate piñones and went out to fetch them. . . . People began well in advance the preparations. Some traveled with carts to live for days in the mountains to gather their piñones, and the whole family was there. It was a very nice life![1]

Mapuche people who live relatively close to Pewen landscapes have, for centuries, temporarily moved to highland forests during the summer and fall with enough provisions to stay a week or more.[2] Early fall is the time when families prepare in advance to make a trip up to the mountains to look for *piñones* or *ngüilliu*, the relished and highly nutritional Pewen seeds. In some territories, gathering *piñones* is a not-to-miss social event. These are also instances for collecting other useful medicinal and edible plants only found in highland forests (Figure 2.2).

For some people, this journey is repeated several times prior to the end of the gathering season. In the past, most families would prepare their wooden carts pulled by a pair of oxen and saddle horses (if available) to get to the *pinaladas* (Pewen forest-stands). They traveled for a couple of days to reach their traditional gathering spot and would camp until they had gathered enough *piñones* for the winter. Nowadays, people are often using trucks to get faster to the gathering spots, which generally allows them to carry a larger amount of *piñones*. Some families prefer to travel longer distances to reach more favorable or preferred gathering spots. "They (grandfather and uncles) liked going to the Küra küra because the piñones were bigger and it was cleaner (meaning open vegetation) to gather or to collect."[3] These gathering trips were also instances for learning and interacting with each other. "*Piñonear* (the act of gathering *piñones*) was a social act because there you met other neighbors who were also hanging around with their families, and they made a fire and elders would sit up most of the night chatting among themselves."[4] The *ngüilliu* is generally considered a 'sacred food' and it is still gathered and consumed with respect to the ancestors and spirits of nature, by asking permission before collecting seeds and praying 'rogativas' before starting the *piñoneo* (Herrmann 2005; Ibarra et al. 2020a). Some community members remember that, in the past, elders would tell stories (*epew*) or historic narratives (*ngütram*) as a way of teaching children about life and especially about how to behave in these forests, teaching philosophies of respect and values for other life-forms. Community gatherings, such as trips to gather *piñones*, were seen as times to reinforce these values. Stories, often interlaced with personal experiences, all impart ecological knowledge as well as cultural perceptions of resources for the benefit of people, to encourage appreciation and also avoid over-extraction (Cruikshank 1990; Turner et al. 2003; Berkes 2012). These traditional modes of communicating, teaching, and learning are tremendously valuable in learning about and practicing sustainable harvesting of wild edible plants (Turner 2000; Turner et al. 2011). They are also instances of nurturing biocultural memory, validating Indigenous epistemologies, and sharing knowledge (Iseke 2014). Stories that our research participants recount always reflect how important these gathering expeditions used to be for the family and community as a whole. They expressed how significant these expeditions were not only for social cohesion and knowledge transmission, but also for securing families with a food of high nutritional value for the winter, when provisions were scarcer.

We have reported that the historical grabbing of Mapuche land by the Chilean state and foreign settlers, the lack of access to forests that used to be traditional gathering grounds by increasing land privatization, and formal schooling are key historical and contemporary drivers of biocultural homogenization of the local food system and the disruption of knowledge transmission associated with the Pewen and other plants (Barreau et al. 2016, 2019). As the gathering of seeds of the Pewen is at the center of human agency and local social fabrics, these processes should be critically analyzed not only in terms of subsistence, but along many dimensions of culture such as identity, social dynamics, institutions, health, and cosmology (Figure 2.3). Mapuche people who relate to the Pewen are, however, not passive recipients of these drivers of change. They are active agents in the processes of generating activities of community resistance, education, and conservation (e.g., "*Mesa Pewenche por la Araucaria*" of *Trawün Lonko*

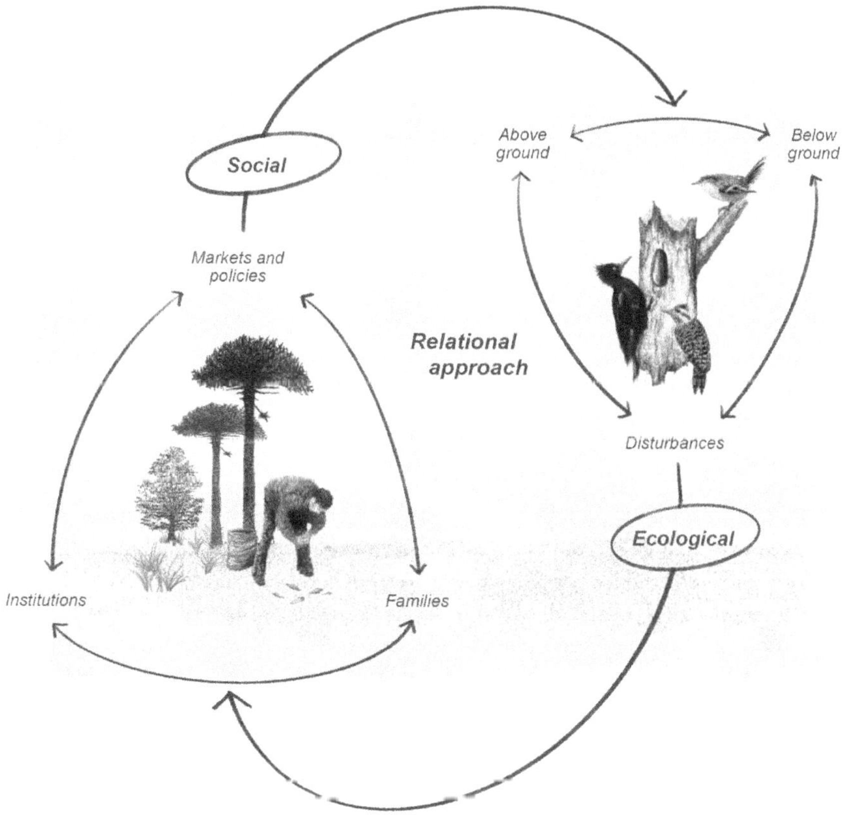

Figure 2.3 A first approach has examined biological entities and their interactions with below- and above-ground fabrics in Pewen forests, examining the role of disturbances shaping ecological dynamics of trees and associated biodiversity. A second approach has explored Mapuche-Pewenche social fabrics linked to the Pewen, which are constrained by historical and contemporary drivers of change such as markets and policies. We propose a third perspective, a relational approach, which highlights unfolding and sympoietic social-ecological relations, overcoming both the ecological/social and the biological entity/human agency dichotomies.

Source: Illustration by M. Medina.

in Lonquimay); creating new market opportunities; and reinforcing spiritual aspects of seed gathering and community bonding. We have reported that local practices related to the gathering of Pewen seeds show the intrinsic and material importance of this seed for both Mapuche people and for long-term non-Mapuche settlers. These practices include the gathering, through the bulk sale of the seeds, to the commercialization of elaborated products (with added value) based on seeds, and a growing offer of tourism experiences that integrate Pewen forests, culture, and local landscapes (Cortés et al. 2019). Therefore, gathering, uses, and commercialization processes associated with the Pewen seeds have at least two main components that establish a continuous dialogue: an economic component, in the commercial and livelihood vein, and a biocultural memory component, in the spiritual, social, and food sense. Along with the physicality of Pewen landscapes, materials, symbolic meanings, and institutions join to constitute the biocultural memory of the landscape (Barthel et al. 2010). This would constitute the basis to nurture situated knowledge about gathering-related practices and rituals associated with the seeds of the Pewen and other multiple cultivars and biodiversity entities in the southern Andes.

Becoming memory: a relational approach to Pewen social-ecological fabrics

As large as some trees are, it is generally hard for us to think of trees as movable, but they do nevertheless, usually slowly. We are not good at thinking about 'movement'; our instinctive skills favor the static, the separate, and the self-contained in its determined subordinating of flux, change, and transformation (Chia 1999). Pewen biocultural landscapes do not consist of static components, but of unpredictable and unfolding 'sympoietic relations' (sympoiesis meaning 'making-with' Haraway 2016). Pewen and its seeds, other plants, wildlife, fungi, soil, climate, people, institutions, the market, and other entities/agents are inseparable from their temporal and spatial dynamic contexts. These entities/agents co-produce each other in complex and sympoietic social-ecological fabrics that vary across space (from the seed to the country and beyond) and time (building on and revisiting the past; Figure 2.2; Darnhofer et al. 2016). Thus, the Pewen, as it is currently, is but a stable moment in a continuing process of 'becoming' (Chia 1999).

Returning again to the example of the Austral parakeet, we can depict how social and ecological fabrics, which were previously allocated to their respective conceptual 'boxes' and posted to either the 'natural sciences' or the 'social sciences,' are inextricably intertwined (Figure 2.3). Mapuche people who gather seeds usually consider the presence of flocks of parakeets on a tree as an indicator of a highly productive Pewen. When people approach these specific trees to collect *piñones*, parakeets escape and leave partially eaten seeds behind, which can still germinate, but people avoid harvesting seeds munched by these birds. The escape from human gathering promoted by the parakeet eating behavior is important for maintaining the population viability of Pewen forests (Speziale et al. 2018). In a different example, the largest woodpecker in South America, the Magellanic woodpecker (*Campephilus magellanicus*), is considered a keystone habitat facilitator for several co-occurring species in nestwebs (Altamirano et al. 2017). Simultaneously, this species is identified by Mapuche people as a reliable weather indicator. According to local participants, Woodpeckers descend from Pewen high-elevation forests when snow, storms, or heavy rain are imminent, advising people to gather wood and prepare themselves to face these harsh weather conditions. Using this bird as a weather forecasting method provides Mapuche farmers with a means for reducing uncertainty while improving decisions in resource management (Dayer et al. 2020).

By eating Pewen seeds or *piñones*, Pewenche become the 'People of the Pewen.' However, a long history of seed collection, dispersion, and planting indicates a deeper relationship between Mapuche-Pewenche and Pewen forest ecology in volcanic lands of the southern Andes. In an ethnographic account, Rozzi (2018) describes that Pewen seeds contain methionine, which is an essential amino acid that cannot be synthesized by the human body. Interestingly, a scientific biogeochemical perspective offers an ecosystem examination of these habits that concurs with the Mapuche worldview. This amino acid contains sulfur in its molecules. Sulfur input to the biogeochemical cycle from volcanoes is transported by the wind and the water of the streams to the soil, where the microflora conduct processes of oxidation and reduction, permitting sulfur to be absorbed by the roots of the Pewen. Therefore, when Pewenche eat the seeds of the Pewen, they also eat the sulfurous rocks and ashes of the volcano. Hence, the Pewenche not only are 'People of the Pewen,' but also Mapuche, 'People of the Land' (Rozzi 2018). This historical people-tree relationship has been based on respect and care. For example, not all cones, neither all trees, are selected for harvesting, and sites with higher yields are favored (Herrmann 2005, 2006). Communities often exclude livestock from their gathering grounds. Also, unsustainable gathering practices like cutting green cones or branches to get down the cones, are recognized and rejected by communities (Cortés et al. 2019). Furthermore, Mapuche-Pewenche have historically helped Pewen forests to thrive by protecting them from fires and logging companies. The case of the Quinquen community in Lonquimay is emblematic, in which Mapuche-Pewenche people of Quinquen allied with environmentalist organizations to reclaim their lands from a forest company that was exploiting Pewen in their historical gathering territories in late 1980s. Moreover, they pressured the government to effectively reinstate the nation-wide protection of the National Monument status of Pewen (Bengoa 1992). Many Pewen trees are currently in their ancestral forested lands because of their interaction with the Mapuche-Pewenche, who have historically dispersed seeds, planted tree saplings, or impeded habitat destruction. These social-ecological interactions nurture the biocultural memory of Pewen human-tree landscapes and have had implications for evolutionary and ecological processes (Herrmann 2006; Reis et al. 2014; Speziale et al. 2018).

Mapuche-Pewenche regard Pewen trees as their protectors. The Pewen allowed the survival of Pewenche in the harsh weather of the Andes mountains, especially during the European and Chilean settler invasion: "[Piñón] is the vital element of survival and of living and resistance of the Mapuche people, and the resistance and survival have to do with the history of how our ancestors came to this territory to take refuge, and to live from the resources that Mother Nature provided. Therefore, that is the teaching that we have of how we value this product and how we value this tree." These memories are still shared today by Mapuche-Pewenche elders, and they remind younger people about the importance of Pewen, and the importance of respect toward this sacred tree. The protector role of the Pewen is part of Mapuche-Pewenche biocultural memory, and it is the basis of a sympoietic relationship between Pewen trees and people.

Final remarks

The Mapuche poet Leonel Lienlaf (1989) recalled that "I became a tree because the tree was my life," embracing the notion that Pewen social-ecological fabrics are never stable but dynamic complex systems that need to be nurtured and performed every day through biocultural memory (Ibarra et al. 2021). These coupled, complex, and sympoietic social-ecological fabrics are nourished with possibilities that can be enriched with novel meanings and understandings. In our own experience working for over a decade in the southern Andes, this implies a radical reconsideration of both how we think about being human and relations with the various entities that make up Pewen landscapes (Murdoch 1997).

Understanding the properties emerging from social-ecological entanglements, such as the biocultural memory of the system, is politically relevant. Negotiations and decisions on social-ecological systems, which affect local human and non-human lives, are based on the understanding of system properties. Neglecting the history of lived interactions and shared experiences, from where the biocultural memory of the system emerges (Toledo and Barrera-Bassols 2008), can reinforce social inequalities affecting Indigenous Peoples' livelihoods, trigger conflicts and, more broadly, reproduce colonialism through an environmental management under preservation approaches (Ludwig 2016; Aedo et al. 2017; Muller et al. 2019). Mapuche-Pewenche and the Pewen have an entangled history that has shaped the becoming of both. Their shared social-ecological experiences nurture a biocultural memory that plays an important role in the present and future life of trees and people. They have protected each other, enabling their survival, and their current sympoietic interrelation and mutual caring is based on this heritage.

In landscapes beyond our case in the southern Andes, how the biocultural memory of the system unfolds, with trees (and their seeds) as repositories of this memory, depends on the history of social-ecological relations, on how local people, institutions, biodiversity, and other actors are constituted through these relations, and how these relations project diverse futures (i.e., the "future memory of the system"; Nazarea 2006; Ibarra et al. 2021). These continuous unfolding materials and symbolic relations will be influenced by events or processes occurring beyond the situated people-tree localities that will constrain some development while making possible other means of 'becoming memory.'

Acknowledgements

This chapter is dedicated in memory of lonko Gumersindo Ayelef Ñanquiñanco and professor Manuel Gedda Ortiz, two wise and humble large trees of Pewen biocultural landscapes. We are grateful to the local and Mapuche community members for their generosity and friendship throughout these years, especially to the lonkos Juan Huilipan, Amador Lefin, and Gumersindo Ayelef. We thank M. Medina for preparing the illustrations for Figures 2.2 and 2.3.

The studies integrated here have been supported by different funding sources, including the ISE Darrell Posey Fellowship, the Namkoong Family Fellowship, the Rufford Small Grants Foundation, the Center for Intercultural and Indigenous Research – CIIR (ANID/FONDAP/15110006), the Cape Horn International Center for Global Change Studies and Biocultural Conservation – CHIC (ANID PIA/BASAL PFB210018), the Center of Applied Ecology and Sustainability – CAPES (ANID PIA/BASAL FB0002), ANID/ REDES (190033), the ANID/Millennium Science Initiative – CESIEP Code NCS13_004, ANID/ FONDECYT Regular 1200291, and ANID/FONDECYT Regular 1181575.

Notes

1 *[porque antes todos comían piñones y los iban a buscar. La gente se preparaba con tiempo. Algunos iban con carros a vivir sus cuatro días en la cordillera para sacar sus piñones y la familia entera. ¡Esa era una vida muy bonita!*] (E ♀6).
2 Only the Pewenche, a subgroup of the Mapuche people, live in between the Pewen trees and they have historically depended on Pewen seeds for their livelihoods as their main staple food.
3 [*A ellos (abuelos y tíos) les gustaba ir al Küra küra porque eran más grandes los piñones y era más limpio para recolectar o recoger*] (D ♀ 1).
4 [*Piñonear era un acto social porque allá se encontraba con otro vecino que también andaba con toda su familia y ahí hacían fogata y se amanecían conversando los viejitos*] (E ♀ 6).

References

Aedo, M. P., S. Peredo, and C. Schaeffer. 2017. From an Essential Being to an Actor's Becoming: Political Ecology Transformational Learning Experiences in Adult Education. *Environmental Education Research*: 1–13. https://doi.org/10.1080/13504622.2017.1408057.

Altamirano, T. A., J. T. Ibarra, K. Martin, and C. Bonacic. 2017. The Conservation Value of Tree Decay Processes as a Key Driver Structuring Tree Cavity Nest Webs in South American Temperate Rainforests. *Biodiversity and Conservation* 26: 2453–2472. https://doi.org/10.1007/s10531-017-1369-x.

Armesto, J. J., D. Manuschevich, A. Mora, C. Smith-Ramirez, R. Rozzi, A. M. Abarzúa, and P. A. Marquet. 2010. From the Holocene to the Anthropocene: A Historical Framework for Land Cover Change in Southwestern South America in the Past 15,000 Years. *Land Use Policy* 27: 148–160. https://doi.org/10.1016/j.landusepol.2009.07.006.

Barreau, A., J. T. Ibarra, F. S. Wyndham, and R. A. Kozak. 2019. Shifts in Mapuche Food Systems in Southern Andean Forest Landscapes: Historical Processes and Current Trends of Biocultural Homogenization. *Mountain Research and Development* 39: 12–23. https://doi.org/10.1659/MRD-JOURNAL-D-18-00015.1.

Barreau, A., J. T. Ibarra, F. S. Wyndham, A. Rojas, and R. A. Kozak. 2016. How Can We Teach Our Children If We Cannot Access the Forest? Generational Change in Mapuche Knowledge of Wild Edible Plants in Andean Temperate Ecosystems of Chile. *Journal of Ethnobiology* 36: 412–432. https://doi.org/10.2993/0278-0771-36.2.412.

Barthel, S., C. Folke, and J. Colding. 2010. Social-Ecological Memory in Urban Gardens: Retaining the Capacity for Management of Ecosystem Services. *Global Environmental Change* 20: 255–265. https://doi.org/10.1016/j.gloenvcha.2010.01.001.

Bengoa, J. 1992. *Quinquén: 100 Años de Historia Pehuenche*. Ediciones ChileAmérica, Santiago, Chile.

Berkes, F. 2012. *Sacred Ecology*, 3rd ed. Routledge, New York.

Blanc, L. A., and J. R. Walters. 2007. Cavity-Nesting Community Webs as Predictive Tools: Where Do We Go from Here? *Journal of Ornithology* 148: 417–423. https://doi.org/10.1007/s10336-007-0232-3.

Briffa, K. R. 2000. Annual Climate Variability in the Holocene: Interpreting the Message of Ancient Trees. *Quaternary Science Reviews* 19: 87–105. https://doi.org/10.1016/S0277-3791(99)00056-6.

Carteron, A., M. Beigas, S. Joly, B. L. Turner, and E. Laliberté. 2020. Temperate Forests Dominated by Arbuscular or Ectomycorrhizal Fungi Are Characterized by Strong Shifts from Saprotrophic to Mycorrhizal Fungi with Increasing Soil Depth. *Microbial Ecology*: 1–14.

Caviedes, J., and J. T. Ibarra. 2017. Influence of Anthropogenic Disturbances on Stand Structural Complexity in Andean Temperate Forests: Implications for Managing Key Habitat for Biodiversity. *PLoS ONE* 12: e0169450. https://doi.org/10.1371/journal.pone.0169450.

Chia, R. 1999. A "Rhizomic" Model of Organizational Change and Transformation: Perspective from a Metaphysics of Change. *British Journal of Management* 10: 209–227. https://doi.org/10.1111/1467-8551.00128.

Cockle, K. L., J. T. Ibarra, T. A. Altamirano, and K. Martin. 2019. Interspecific Networks of Cavity-Nesting Vertebrates Reveal a Critical Role of Broadleaf Trees in Endangered Araucaria Mixed Forests of South America. *Biodiversity and Conservation* 28: 3371–3386. https://doi.org/10.1007/s10531-019-01826-4.

Cortés, J., I. Ugalde, J. Caviedes, and J. T. Ibarra. 2019. Semillas de Montaña: Recolección, Usos y Comercialización Del Piñón de La Araucaria (*Araucaria Araucana*) Por Comunidades Mapuche-Pewenche Del Sur de Los Andes. *Pirineos* 174: 048. https://doi.org/10.3989/pirineos.2019.174008.

Cruikshank, J. 1990. Getting the Words Right: Perspectives on Naming and Places in Athapaskan Oral History. *Arctic Anthropology* 27: 52–65.

Darnhofer, I., C. Lamine, A. Strauss, and M. Navarrete. 2016. The Resilience of Family Farms: Towards a Relational Approach. *Journal of Rural Studies* 44: 111–122. https://doi.org/10.1016/j.jrurstud.2016.01.013.

Dayer, A. A., E. A. Silva-Rodríguez, S. Albert, M. Chapman, B. Zukowski, J. T. Ibarra, G. Gifford, A. Echeverri, A. Martínez-Salinas, and C. Sepúlveda-Luque. 2020. Applying Conservation Social Science to Study the Human Dimensions of Neotropical Bird Conservation. *Condor* 122: 1–15. https://doi.org/10.1093/condor/duaa021.

de Vos, A., R. Biggs, and R. Preiser. 2019. Methods for Understanding Social-Ecological Systems: A Review of Place-Based Studies. *Ecology and Society* 24: 1–24. https://doi.org/10.5751/ES-11236-240416.

Díaz, S., and T. Kitzberger. 2013. Nest Habitat Selection by the Austral Parakeet in North-Western Patagonia. *Austral Ecology* 38: 268–278.

Diehl, P., and S. Fontenla. 2010. Arbuscular Mycorrhizal Infection in Two Morphological Root Types of Araucaria Araucana (Molina) K. Koch. *Revista Argentina de Microbiología* 42: 133–137.

Donoso, C. 1993. *Bosques Templados the Chile y Argentina: Variación, Estructura y Dinámica.* Editorial Universitaria, Santiago, Chile.

Filotas, E., L. Parrott, P. J. Burton, R. L. Chazdon, K. D. Coates, L. Coll, S. Haeussler, K. Martin, S. Nocentini, K. J. Puettmann, F. E. Putz, S. W. Simard, and C. Messier. 2014. Viewing Forests through the Lens of Complex Systems Science. *Ecosphere* 5: 1–23.

Godoy, R., and C. Marín. 2019. Mycorrhizal Studies in Temperate Rainforests of Southern Chile. In *Mycorrhizal Fungi in South America*, edited by M. C. Pagano and M. A. Lugo, pp. 315–341. Springer, Cham, Switzerland. https://doi.org/10.1007/978-3-030-15228-4_16.

González, M. E., T. T. Veblen, and J. S. Sibold. 2005. Fire History of Araucaria-Nothofagus Forests in Villarrica National Park, Chile. *Journal of Biogeography* 32: 1187–1202. https://doi.org/10.1111/j.1365-2699.2005.01262.x.

Hageneder, F. 2009. *El Legado de Los Árboles: Historia, Cultura y Simbolismo*. Ediciones Columba, Santiago, Chile.

Haraway, D. J. 2016. *Staying with the Trouble: Making Kin in the Chthulucene*. Duke University Press, Durham.

Herrmann, T. M. 2005. Knowledge, Values, Uses and Management of the Araucaria Araucana Forest by the Indigenous Mapuche Pewenche People: A Basis for Collaborative Natural Resource Management in Southern Chile. *Natural Resources Forum* 29: 120–134. https://doi.org/10.1111/j.1477-8947.2005.00121.x.

Herrmann, T. M. 2006. Indigenous Knowledge and Management of Araucaria Araucana Forest in the Chilean Andes: Implications for Native Forest Conservation. *Biodiversity and Conservation* 15: 647–662. https://doi.org/10.1007/s10531-005-2092-6.

Holmgren, M., and M. Scheffer. 2017. To Tree or Not to Tree: Cultural Views from Ancient Romans to Modern Ecologists. *Ecosystems* 20: 62–68. https://doi.org/10.1007/s10021-016-0052-5.

Huiliñir-Curío, V. 2010. El Rol de Las Veranadas En El Territorio Pehuenche de Alto Bio Bio. Sector Lonquimay, IX Región. *Revista Geográfica Despertando Latitudes* 2: 17–24.

Ibarra, J. T., T. A. Altamirano, N. Gálvez, I. Rojas, J. Laker, and C. Bonacic. 2010. Avifauna de Los Bosques Templados de Araucaria Araucana Del Sur de Chile. *Ecología Austral* 20: 33–45.

Ibarra, J. T., A. Barreau, J. Caviedes, N. Pessa, J. Valenzuela, S. Navarro-Manquilef, C. Monterrubio-Solís, A. Ried, and J. C. Pizarro. 2021. Listening to Elders: Birds and Forests as Intergenerational Links for Nurturing Biocultural Memory in the Southern Andes. In *Transnational Children and Youth: Experiences of Nature and Place, Culture and Care across the Americas*, edited by V. Derr and Y. Corona, pp. 160–175, 1st ed. Routledge, Abingdon, UK.

Ibarra, J. T., A. Barreau, C. Marchant, J. A. González, M. Oliva, M. E. Donoso-Correa, B. Antaki, C. Monterrubio-Solís, and F. O. Sarmiento. 2020a. Montology: An Integrative Understanding of Mountain Foodscapes for Strengthening Food Sovereignty in the Andes. In *The Elgar Companion to Geography, Transdisciplinarity and Sustainability*, edited by F. O. Sarmiento and L. M. Frolich, pp. 391–405. Edward Elgar Publishing, Cheltenham, UK.

Ibarra, J. T., K. L. Cockle, T. A. Altamirano, Y. Van der Hoek, S. Simard, C. Bonacic, and K. Martin. 2020b. Nurturing Resilient Forest Biodiversity: Nest Webs as Complex Adaptive Systems. *Ecology and Society* 25: 27. https://doi.org/10.5751/ES-11590–250227.

Ibarra, J. T., N. Gálvez, T. A. Altamirano, J. Caviedes, I. M. Rojas, C. Bonacic, and K. Martin. 2017. Seasonal Dynamics of Avian Guilds inside and Outside Core Protected Areas in an Andean Biosphere Reserve of Southern Chile. *Bird Study* 64: 410–420. https://doi.org/10.1080/00063657.2017.1368447.

Ibarra, J. T., and K. Martin. 2015. Biotic Homogenization: Loss of Avian Functional Richness and Habitat Specialists in Disturbed Andean Temperate Forests. *Biological Conservation* 192: 418–427. https://doi.org/10.1016/j.biocon.2015.11.008.

Ibarra, J. T., K. Martin, M. C. Drever, and G. Vergara. 2014. Occurrence Patterns and Niche Relationships of Sympatric Owls in South American Temperate Forests: A Multi-Scale Approach. *Forest Ecology and Management* 331: 281–291.

Ibarra, J. T., F. J. Novoa, H. Jaillard, and T. A. Altamirano. 2020c. Large Trees and Decay: Suppliers of a Keystone Resource for Cavity-Using Wildlife in Old-Growth and Secondary Andean Temperate Forests. *Austral Ecology* 45: 1135–1144. https://doi.org/10.1111/aec.12943.

Infante, J., F. J. Novoa, J. T. Ibarra, D. J. Melnick, K. L. Griffin & C. Bonacic. 2021. Altered fire regimes modify lizard communities in globally endangered Araucaria forests of the southern Andes. *Scientific Reports* 11: 22709.

Iseke, J. 2014. Indigenous Storytelling as Research. *International Review of Qualitative Research* 6: 559–577. https://doi.org/10.1525/irqr.2013.6.4.559.559.

IUCN. 2021. *IUCN Red List of Threatened Species*. Version 2021–1. Available at www.iucnredlist.org.

Jones, O., and P. Cloke. 2002. *Tree Cultures: The Place of Trees and Trees in Their Place. Anthropologica*. Berg, Oxford, UK. https://doi.org/10.2307/25606157.

Jusino, M. A., D. L. Lindner, M. T. Banik, K. R. Rose, and J. R. Walters. 2016. Experimental Evidence of a Symbiosis between Red-Cockaded Woodpeckers and Fungi. *Proceedings of the Royal Society B: Biological Sciences* 283: 20160106. https://doi.org/10.1098/rspb.2016.0106.

Kohn, E. 2013. *How Forests Think? Toward an Anthropology beyond the Human*. University of California Press, Berkeley, CA.

Lienlaf, L. 1989. *Se Ha Despertado El Ave de Mi Corazón*. Editorial Universitaria, Santiago, Chile.

Lindenmayer, D. B., and W. F. Laurance. 2017. The Ecology, Distribution, Conservation and Management of Large Old Trees. *Biological Reviews* 92: 1434–1458. https://doi.org/10.1111/brv.12290.

Ludwig, D. 2016. Overlapping Ontologies and Indigenous Knowledge: From Integration to Ontological Self-Determination. *Studies in History and Philosophy of Science* 59: 36–45.

Maestre, F. T., J. L. Quero, N. J. Gotelli, A. Escudero, V. Ochoa, M. Delgado-Baquerizo, M. García-Gómez, M. A. Bowker, S. Soliveres, C. Escolar, P. García-Palacios, M. Berdugo, E. Valencia, B. Gozalo, A. Gallardo, L. Aguilera, T. Arredondo, J. Blones, B. Boeken, D. Bran, A. a Conceição, O. Cabrera, M. Chaieb, M. Derak, D. J. Eldridge, C. I. Espinosa, A. Florentino, J. Gaitán, M. G. Gatica, W. Ghiloufi, S. Gómez-González, J. R. Gutiérrez, R. M. Hernández, X. Huang, E. Huber-Sannwald, M. Jankju, M. Miriti, J. Monerris, R. L. Mau, E. Morici, K. Naseri, A. Ospina, V. Polo, A. Prina, E. Pucheta, D. a Ramírez-Collantes, R. Romão, M. Tighe, C. Torres-Díaz, J. Val, J. P. Veiga, D. Wang, and E. Zaady. 2012. Plant Species Richness and Ecosystem Multifunctionality in Global Drylands. *Science* 335: 214–218. https://doi.org/10.1126/science.1215442.

Martin, K., K. E. H. Aitken, and K. L. Wiebe. 2004. Nest Sites and Nest Webs for Cavity-Nesting Communities in Interior British Columbia, Canada: Nest Characteristics and Niche Partitioning. *Condor* 106: 5–19.

Martin, K., and J. M. Eadie. 1999. Nest Webs: A Community-Wide Approach to the Management and Conservation of Cavity-Nesting Forest Birds. *Forest Ecology and Management* 115: 243–257.

Molina, R., and T. R. Horton. 2015. Mycorrhiza Specificity: Its Role in the Development and Function of Common Mycelial Networks. In *Mycorrhizal Networks. Ecological Studies (Analysis and Synthesis)*, edited by T. R. Horton, pp. 1–39. Springer, Dordrecht, Netherlands.

Muller, S., S. Hemming, and D. Rigney. 2019. Indigenous Sovereignties: Relational Ontologies and Environmental Management. *Geographical Research* 57: 399–410.

Murdoch, J. 1997. Inhuman/Nonhuman/Human: Actor-Network Theory and the Prospects for a Nondualistic and Symmetrical Perspective on Nature and Society. *Environment and Planning D: Society and Space* 15: 731–756. https://doi.org/10.1068/d150731.

Nazarea, V. D. 2005. *Cultural Memory and Biodiversity*. The University of Arizona Press, Tucson, AZ.

Nazarea, V. D. 2006. Local Knowledge and Memory in Biodiversity Conservation. *Annual Review of Anthropology* 35: 317–335. https://doi.org/10.1146/annurev.anthro.35.081705.123252.

Novoa, F., T. A. Altamirano, C. Bonacic, K. Martin & J. T. Ibarra. 2021. Fire regimes shape biodiversity: responses of avian guilds to burned forests in Andean temperate ecosystems of southern Chile. *Avian Conservation and Ecology* 16(2): 22.

Reis, M. S. dos, A. Ladio, and N. Peroni. 2014. Landscapes with Araucaria in South America: Evidence for a Cultural Dimension. *Ecology and Society* 19: 43.

Rozzi, R. 2018. Biocultural Ethics: From Biocultural Homogenization toward Biocultural Conservation. In *From Biocultural Homogenization toward Biocultural Conservation*, edited by R. Rozzi, R. H. May Jr, F. S. Chapin III, F. Massardo, M. C. Gavin, I. Klaver, A. Pauchard, M. A. Nuñez, and D. Simberloff, pp. 9–32. Springer, New York. https://doi.org/10.1007/978-3-319-99513-7.

Shepherd, J. D., R. S. Ditgen, and J. Sanguinetti. 2008. Araucaria Araucana and the Austral Parakeet: Predispersal Seed Predation on a Masting Species. *Revista Chilena de Historia Natural* 81: 395–401. https://doi.org/10.4067/S0716-078X2008000300008.

Simard, S. W. 2018. Mycorrhizal Networks Facilitate Tree Communication, Learning, and Memory. In *Memory and Learning in Plants*, edited by F. Baluška, M. Gagliano, and G. Witzany, pp. 191–213. Springer, Cham, Switzerland.

Simard, S. W., K. Martin, A. Vyse, and L. Bruce. 2013. Meta-Networks of Fungi, Fauna and Flora as Agents of Complex Adaptive Systems. In *Managing Forests as Complex Adaptive Systems: Building Resilience to the Challenge of Global Change*, edited by C. Messier, K. J. Puettmann, and K. D. Coates, pp. 133–164. Routledge, New York.

Simard, S. W., D. A. Perry, M. D. Jones, D. D. Myrold, D. M. Durall, and R. Molina. 1997. Net Transfer of Carbon between Ectomycorrhizal Tree Species in the Field. *Nature* 388: 579–582.

Smith, S. E., and D. J. Read. 2010. *Mycorrhizal Symbiosis*. Academic Press, New York.

Speziale, K. L., S. A. Lambertucci, G. Gleiser, J. L. Tella, F. Hiraldo, and M. A. Aizen. 2018. An Overlooked Plant-Parakeet Mutualism Counteracts Human Overharvesting on an Endangered Tree. *Royal Society Open Science* 5: 171456. https://doi.org/10.1098/rsos.171456.

Talbot, J. M., S. D. Allison, and K. K. Treseder. 2008. Decomposers in Disguise: Mycorrhizal Fungi as Regulators of Soil C Dynamics in Ecosystems under Global Change. *Functional Ecology* 22: 955–963.

Teste, F. P., M. D. Jones, and I. A. Dickie. 2020. Dual-Mycorrhizal Plants: Their Ecology and Relevance. *New Phytologist* 225: 1835–1851.

Toledo, V. M., and N. Barrera-Bassols. 2008. *La Memoria Biocultural: La Importancia Ecológica de Las Sabidurías Tradicionales*. Icaria Editorial, Barcelona, Catalunya.

Turner, N. J. 2000. "Keeping It Living": Applications and Relevance of Traditional Plant Management in British Columbia to Sustainable Harvesting of Non-Timber Forest Products. In *NTFP Conference Proceedings*, pp. 66–77. USDA Forest Service, Madison, Wisconsin.

Turner, N. J., I. J. Davidson-Hunt, and M. O'Flaherty. 2003. Living on the Edge: Ecological and Cultural Edges as Sources of Diversity for Social-Ecological Resilience. *Human Ecology* 31: 439–461.

Turner, N. J., Ł. J. Łuczaj, P. Migliorini, A. Pieroni, A. L. Dreon, L. E. Sacchetti, and M. G. Paoletti. 2011. Edible and Tended Wild Plants, Traditional Ecological Knowledge and Agroecology. *Critical Reviews in Plant Sciences* 30: 198–225. https://doi.org/10.1080/07352689.2011.554492.

Watkins, C. 2016. *Trees, Woods and Forests: A Social and Cultural History*. Reaktion Books, London.

3 White pine

The tree that sparked peace, revolution, and insurrection

Patricia Shanley

Introduction

Upon arrival in the New World, British colonists beheld towering, 200-foott pine trees piercing the sky. Without these massive trees, a new nation may never have been born. Strong and light weight, white pines became the premier mast-making tree worldwide with Britain, Portugal, and Madagascar competing to procure them. Revenues from white pine masts underwrote the Revolutionary War and the trees helped to build a new nation (Vietz 2018).

White pine adorned the first flag of the revolutionary forces, graced the shilling, built bridges, and furnished tea to keep scurvy at bay (Rutkow 2012). When the British Crown ordered the most valuable pines to be branded for the Royal Navy, fury arose among hard scrabble farmers. White pine riots ensued. Psychologically and economically, white pines fortified the fledgling colonies against the repression of the British, helping to catalyze the American Revolution (Peattie 1950).

Lesser known is that the white pine had already played a leading role in the founding of a prior nation – The Haudenosaunee or Iroquois Confederation. Prior to European contact, the white pine was central to Native American governance, having helped give birth to a sophisticated system of representation with peace at its core (Wallace 1946). The white pine was the tree under which weapons were laid down, in exchange for an accord based on reconciliation and good will. When the Europeans arrived, challenges over land ensued and war commenced again.

Once the colonists vanquished the British, and drove out Indigenous communities, the young colonies were compelled to devise a new system of governance. Ironically, key elements of the progressive governance system of the Iroquois directly helped to lay the foundation of the US Constitution including a balance of power, two branches of government, and designated procedures for passing laws.

Fatefully, 245 years after the founding of the United States, the white pine flag has been resurrected. A potent political banner, it was revived by insurrectionists on January 6th, 2021 as a symbol of overthrowing tyrannical rule. The flag that was once proudly embraced by colonial settlers in defiance of the British King, was waved by self-proclaimed "patriots" who stormed the US

DOI: 10.4324/9780429320897-4

Capitol. In this case, the white pine symbol was used to help catalyze an attempt to overthrow a democratically held election.

This chapter traces the livelihood uses and enduring symbolic value of the white pine from about 1,000 years ago to the present. It begins by recalling Indigenous management, use, and symbolism of the tree, followed by a description of its use and meaning to early colonists and its role in helping to spark the Revolutionary War. Next, it examines recent appropriation of the symbol by contemporary political groups. Subsequently, the present chasm between people and trees is considered, and an example focusing on the white pine is presented, describing an initiative to boost interest in and respect for trees. During the twin crises of biodiversity loss and climate change, deeper understanding of the profound role of trees and forests in everyday life may help ameliorate the widespread disregard displayed toward trees.

Background

The topic of this book, *Tree Folk Value: Folk Value and Biocultural Conservation*, raises fundamental questions. What relationship do people have with trees today? In an urbanized, technologically driven society, do many people, other than forest-reliant communities, gardeners, farmers, and/or landscapers, see, consider, or interact with trees? Or is such a connection to trees a vestige of the past?

Another issue is the relationship between researchers who study trees, and the communities within which they work. Culture and the human connection to trees are not topics natural scientists are generally encouraged, nor trained, to study. Researchers are taught to measure and analyze the ecosystem services, ecology, or the monetary worth of trees – their timber, exudates, fruits, medicines, game attracting potential, pollinators, markets and, possibly, their recreational or therapeutic values. The customary use of economics or environmental services as a primary methodological lens through which to study trees, however, can blind researchers to other significant factors. The ties that bind people to trees – symbols, songs, stories, recipes, myths, lore, and emotional connections – often remain veiled.

Working as a forest ecologist for two decades with strong-willed urban and suburban youth in forests of New Jersey, USA, and for decades with farmers and hunter-gatherers in the Brazilian Amazon, has led me to understand that what is least frequently examined – cultural and emotional connections – are of primary significance in people's relationships with trees. It is the stories behind the numbers that embed trees in people's lives. And, as the case of the white pine illustrates, it is these stories and experiences that are fading.

For thousands of years, the white pine has been revered by Indigenous tribes in North America as a sacred tree. Native legends, poems, stories, and myths feature the white pine as a central figure (Schroeder 1992). The white pine is also recognized by historians of the American Revolution as having played a pivotal economic and psychological role in the revolutionaries' victory in the war against the British Crown. Presently, most Americans cannot distinguish a pine from a spruce or fir, much less possess knowledge of the position that this once-grand tree had in launching not one, but two, daring experiments in democracy. Understanding the influence of trees and their symbolic value on our minds, actions, and political and cultural landscapes, is particularly timely as the rupture between nature and people grows and peace dwindles.

The cultural and nutritional landscape of Indigenous North America

When British explorers arrived in North America they encountered not an untrammeled wilderness, but a landscape that had been skillfully managed for centuries by Native Americans (Denevan 1992). White pine trees pierced the canopy, being emergent and the tallest trees in the region. White pine were prevalent in the Northeastern forests of the US, sometimes growing as scattered individuals and in other areas as nearly pure stands. Without disturbance, such trees could live to be up to 500 years old.

At the time of European settlement, white pines occupied a range of soil and moisture conditions and were well adapted to disturbance regimes including fire and blowdowns (Abrams and Nowacki 2008). Their distribution includes what is now southern Newfoundland, through New England, New York, Pennsylvania, the Great Lakes region, and along the Appalachian Mountains to northern Georgia. There is substantial evidence that fire was used systematically by Native Americans to create fields, clear undergrowth, diminish rodents, and to enhance hunting by expanding open fields and forage (Denevan 1992; William 2005).

Intermittent burns encouraged recruitment and long-term success of tree species such as oak, hickory, chestnut, and pine. Together with clearing, girdling, and planting, a patchwork of various aged fields and secondary forests resulted, increasing species diversity and creating dense groves of nut trees and edible berry bushes surrounding villages.

The exceptional richness of flora and fauna in North America contributed a vital nutritional foundation for the health of Native Americans, and later, colonists. Native Americans consumed at least 30 tree species and genera, including oak (*Quercus*), hickory (*Carya*) and chestnut (*Castanea*), and walnut (*Juglans*) and pine (*Pinus*) (Abrams and Nowacki 2008). These supplied crucial fats, carbohydrates, vitamins, and nutrients to peoples reliant on wild resources for their sustenance. Pines were used to make tea from their needles, their bark was eaten during periods of famine, and their resin was used as a food flavoring.

The Iroquois peace tree

Complex systems of forest management existed alongside sophisticated systems of Indigenous governance. Rarely encountered in history books or taught in schools, The Haudenosaunee Confederacy (meaning People of the Long House, or Iroquois in French and Six Nations in English) is one of the oldest living participatory democracies on Earth. Based on documentary sources, oral history, and celestial events, the body of law of the United Nations of the

Iroquois was ratified on August 31, 1142, at a site called Gonandaga by the Seneca, now named Victor, New York (Mann and Fields 1997; Hansen 2018).

This Confederation came about as part of a decades-long process of consensus building among formerly warring tribes. Ongoing fighting had eroded and weakened the tribes' social order. A visionary named, Deganawidah (The Peacemaker) and the Chief of the Onondaga tribe, Hiawatha, foresaw a means to bring peace among the nations. Over a period of 25 years, they journeyed to council meetings in each of the Indigenous nations, Mohawk, Oneida, Onondaga, Cayuga, and Seneca to present the Great Law of Peace. Charismatic and convincing, the pair discussed governance systems at successive council meetings, resolving that individual nations would maintain their own leadership, but common causes would be decided in the Grand Council of Chiefs. Central to the law was the democratic concept of consensus, basing resolutions on peace rather than fighting. A female leader of clan mothers was also part of the discussions, ensuring an equal and strong place for women in decision-making (Reilly 2020).

After 25 years of deliberation among the tribes, the Iroquois Confederacy Constitution was ratified in 1142 by the five nations (Mann and Fields 1997). In 1722, the Tuscarora would also join the Iroquois Confederacy, expanding the alliance to include six nations. On the auspicious day in 1142, a solar eclipse occurred, likely heightening the perception of awe among the tribes in the powers of Deganawidah. Under the agreement, tribes agreed to lay down their weapons of war under the roots of a massive white pine tree. The white pine tree pierced the sky to reach the sun, signifying the law of peace. As Wallace (1946) writes, "The Iroquois fed their minds and guided their actions by means of symbols." The branches denoted the shelter and security of union under the law, and the roots the extension of peace to the four corners of the earth, compelling others to take shelter under the branches and embrace harmony. An eagle alighting atop the tree, signified vigilance, a sentry to sound warning if any evil approached through the roots.

The white pine – luring British across the Atlantic

In contrast to the flourishing, highly diverse forests of the Iroquois Federation, the forests of Britain had been decimated by as early as the mid 1500s. In response to the country's rapidly diminishing timber resources, Parliament passed the "Act of Preservation of Woods" in 1543. By 1590, only one eighth of England's original forest remained. The dearth of timber posed a critical challenge as wood from the Baltic states was becoming increasingly expensive and the Royal Navy required wood for masts and shipbuilding to retain command over its colonies worldwide (Vietz 2018)

As early as 1584, British geographer Richard Hakluyt recognized the astonishing timber resource on the other side of the Atlantic Ocean as a potential asset to the Royal Navy. In 1584, he met with Queen Elizabeth I to appeal for a charter to establish British colonies in the New World but was denied. For the next two decades, he collected travel accounts from explorers and published these in a compendium entitled, *The Principal Navigations*. Having 20 years of expertise behind him, in 1605, he met with King James I, who agreed to issue the Charter of Virginia (Rutkow 2012). Hakluyt is a model of perseverance – it took 21 years until his vision became realized.

As opposed to the denuded landscape of Britain, 95% of New England's land area was forested when the colonists arrived, and the massive sizes of the trees were inconceivable. Legend from the time recounts that a squirrel could travel all the way from the East Coast to the Mississippi River without touching ground! Having grown with minimal disturbance for centuries, New World trees offered old, close growth rings which contributed to strength, creating a fine grade of wood, free of defect and knot, and rot resistant (Rutkow 2012).

In contrast to dense, heavy tropical timbers, white pines are tall, light, flexible, and strong – perfect for masts. White pine swiftly became the most highly valued tree in the world for mast building and the most lucrative manufactured export from the colonies. In 1623, the first sawmill was constructed in York, Maine. As early as 1629, one merchant recalled, "There is no better ship timber to be found in the world" (Vietze 2018).

Felling a massive white pine was no small feat. Many trees needed to be cleared in the direction of the felling to create a landing area and great care was needed to ensure that the highly valuable timber did not crack. Bedding to soften the blow was carefully placed along the tree's fall line and snow was helpful in buffering a fall (Manning 1979) (Figure 3.1).

Figure 3.1 New England Masts & the King's Broad Arrow by Samuel F. Manning.

Once the hard work of felling the gigantic trees was complete, an enormous amount of labor and skill was required to haul the timber to a river to reach a sawmill. Incentivized by the superior profits, colonists toiled in the forests, undertaking the colossal task of felling and transporting white pines to be shipped as far away as Madagascar, Spain, Britain, Portugal, and the West Indies. The mighty ships fashioned with these masts became known as, "wooden cathedrals of the oceans."

White pine – sparking a revolution and building a nation

Sales of timber for the shipping industry transformed the rugged colonies from what they experienced as a harsh land into a wealthy trading post (Vietze 2018). Over the next one hundred years, colonists controlled one of the strongest shipping trades outside of the Dutch empire, with some colonial timber barons becoming as rich as Englishmen. White pine wood served as currency, the tree was featured on colonial shillings, and the first flag of revolutionary forces showcased a white pine, with the words "An Appeal to Heaven" above it. Thus began the colonists' separation from Britain.

In 1691, King William III, recognizing that the colonists were usurping not only timber, but also power, money, and masts from the Crown, issued a decree stating, "Notice is given that all White Pine Logs cut and hauled out of the King's Woods . . . will be seized to his Majesty's Use and Trespassers dealt with according to law" (Vietze 2018: 69). The King sent surveyors to brand all white pines greater than 24 inches in diameter with the Kings Broad Arrow (an upside-down arrow). In 1720, a mast tree of 36 inches could fetch a huge sum. For colonists, who had labored tirelessly to carve out a living in a daunting new landscape, the Broad Arrow represented the repressive Crown infiltrating their lives and forests.

Through various means, colonists throughout New England rebelled against the King's decree. The woods of New Hampshire and Maine were famous for Pine Tree Riots and Woodland Rebellions. Many settlers clandestinely continued to fell white pines, some dressing as Indians and working during the night. When encountering the King's men overseeing the white pines, some colonists declared, "Swamp Law," violently beating the King's men, and running them out of their forests and towns. Anger over royal privilege appropriating what they perceived as their own forests infuriated settlers throughout New England, producing numerous skirmishes and riots. In the region that is now Maine and New Hampshire, clashes between the King's men and colonists led to mast tree rebellions in 1734 and 1772. History highlights the Stamp Act of 1765 and the Tea Tax of 1773 as sparking the American Revolution, but the white pine riots helped to catalyze the revolution and then, literally, the building of the new nation (Vietze 2018).

After winning the revolution, colonists again turned to the white pine. The tree was felled for the construction of bridges, barns, railroads, cabins, hobby horses, ship figureheads, shingles, roofs, and bobsleds. North America was literally and figuratively built out of white pine (Rutkow 2012). By the late 1800s, when settlers turned to see what they had wrought, extensive logging to supply demand had wiped out the grand conifers. The magnificent white pines, once perceived as an endless supply of timber, had been extirpated.

The white pine plants seeds of the US Constitution

The legacy of the white pine, however, lived on. The Iroquois Confederacy, founded alongside the White Pine, had brought centuries of peace among formerly warring tribes. As the first thirteen colonies struggled to become the United States of America, the governance system that the white pine had given birth to, offered an unexpected role model. In Europe at the time, there were no participatory democracies from which the delegates could derive ideas and inspiration. In preparation for the Constitutional Convention held in Philadelphia in 1781, John Adams wrote a three-volume handbook examining different types of governing bodies. While Adams included thinkers like John Locke and Montesquieu, he also described elements of enduring governance systems from the Iroquois Confederacy, which many delegates knew through personal experience.

The Federation's constitution was documented and kept alive through designs on a wampum belt. Front and center stood a profile of a white pine, with designs symbolizing concepts of governance familiar to United States citizens today. For example: a balance of power is held between the Iroquois Confederacy and individual tribes; two branches of legislature with procedures for passing laws is designated; members are restricted from holding more than one office; processes to remove leaders within the Confederacy are outlined; and who has the power to declare war is delineated.

In 1988, recognizing that the founding fathers of the United States built directly upon Indigenous land and governing principals, congress passed a resolution, formally acknowledging the influence of the Iroquois Confederacy on the US Constitution. In addition to this recognition, the resolution reaffirmed "the continuing government-to-government relationship between Indian tribes and the United States established in the Constitution" (U.S. Congress 1988).

White pine flag stokes insurrection

Two hundred fifty-five years after the founding of the United States, the symbol of the white pine that stood for a newly founded representative republic is now being waved at insurrectionist rallies. In 1690, John Locke wrote that when an oppressed people "have no appeal on earth, then they have a liberty to appeal to Heaven" (Vietze 2018). This phrase was initially interpreted by colonists as against the Divine Right of the Rule of Kings and was first flown by a six-cruiser squadron under George Washington's command in October 1775. During the last two decades, the White Pine Flag of the Revolution bearing the words, "An Appeal to Heaven" has been appropriated by a range of seemingly incongruent groups. On January 6, 2021, a riotous mob which included some individuals bearing this flag, stormed the US Capitol in an attempt to overturn the results of a democratically held election.

In 1775, the bearers of the flag recognized the species of tree as a white pine and understood how deeply the tree was intertwined with the arduous lives of colonial families. They razed the tree, hauled it, transported it over woods and waters, and sold it to survive. The King's decree to seize mature white pines epitomized the tyranny of royal privilege and oppression and mobilized colonists to break free of the Crown. Today, the symbol of the tree and the inscription, "An Appeal to Heaven," also resonates to mobilize political forces. But in this case, it is used to attempt to overthrow a system of representative governance that the revolutionary forces fought for over 250 years ago. Today, few Americans know what species of tree it is or the role it played in Indigenous or colonial life. Of 100 people interviewed as part of this research in Maine, none recognized the "Appeal to Heaven" flag or had a notion of what it means. In a 2020 *New York Times* article about the Maine State flag, which also bears the profile of a white pine, the species of tree was never mentioned, nor the history of the original white pine flag on which it is based (Boylan 2020).

How did the White Pine Flag of the colonists come to be interpreted as it is today? In the early 21st century, the flag became adopted by religious conservatives in the Republican Party linked with faith-based Christian movements. During the era of Billy Graham in evangelical circles, politics began to be openly infused into religion. Then in 2008, Republican presidential nominee John McCain chose a right-wing populist with evangelical leanings as his running mate. Since then, the line has blurred between politics and religion and faith-fueled politics has been on the rise. One initiative called "An Appeal to Heaven," using the white pine revolutionary flag as its emblem, was founded by a South Carolina preacher who has 200,000 followers on YouTube. The movement has tenants of Christian nationalism, with overtones of white supremacy. Followers include self-described anti-government militias, white nationalist agitators, and Bible toting evangelicals. At rallies, the Proud Boys pray to God while flanked by evangelical Christians holding the ichthys (fish) signs, a popular symbol used by evangelicals to evoke Jesus Christ, God's Son and Savior (Winston 2021).

The white pine flag has been flown at the conservative Political Action Conference in 2017, at gun rights rallies in 2020, neo-fascist events in Ohio in 2021, at the Call to Arms rally in South Carolina, at anti-lockdown events during the COVID-19 pandemic, and the Save America Rally on January 6, 2021. It has also been intentionally positioned behind republican senators and a former US president at conservative rallies. Widespread adoption of the flag by disparate groups bearing other flags including Norse, neo-confederate, and Christian banners, indicate that movements which may formerly have had little in common are ideologically coalescing through the use of adopted symbols (Jhaver 2021). Present-day flag bearers wave it, blithely appropriating the symbol with no knowledge of the original Indigenous meaning of the white pine as signifying peace (Figure 3.2).

Figure 3.2 White pine flag flying alongside the American Flag in front of a church in Southern New Jersey, USA.

Scholars studying the movement warn that the deep roots and widespread adoption of Christian nationalism throughout the US "is a threat to a pluralistic democratic society. . . [and] people go under the guise of Christian symbols and symbolism enact violence against their own country" (Whitehead 2021). Dan Webster, a retired Episcopal pastor, remarked, "I [am] ashamed of Christian leaders these last 4 years whose silence has been compliance." Another commentator, fond of the white pine flag, stated, "Dammit, why is that pine tree flag used by Fascists? That's too similar to the Maine state flag for my liking (and I liked that flag until now)" (Winston 2021).

Disconnecting from trees

In contrast to the colonial era when the lives of families were directly dependent upon forests, and the white pine was widely recognized, people's interaction with trees today is miniscule. In fact, surveys indicate that North Americans spend approximately 90% of their lives in buildings (Klepeis et al. 2001). Urbanization, technology, and an increasingly indoor culture, devoid of contact with nature, contributes to a vast rift between people and trees. Demographics indicate that this rupture is not mending but intensifying. The radical shift to becoming an indoor species raises fundamental questions posed at the beginning of this chapter: in an urbanized, technologically driven society, what impact will a lack of experience with trees have on culture and our relationship with and actions toward nature?

These questions are particularly important to address as the Earth faces the twin dilemmas of a warming planet and species extinction. These crises cannot be met by policies alone but must also be met by the daily actions of billions of individuals, locally. Research demonstrates that inculcating a life-long connection to nature occurs prior to 12 years of age. How will the next generation attain a sufficient affinity with trees and nature to take the actions needed to mitigate the irreversible impacts of climate change and the staggering loss of biodiversity and forests?

Reconnecting to trees: stories, lore, and action in woods

In many US schools today, it is against the rules to climb a tree or to throw a snowball. No wonder so many kids are depressed. Considering the cliff-hanging environmental cataclysm society is surging toward, where and how children spend their time, and what they learn, needs to be fundamentally reconsidered. Since children pass the bulk of their young lives in school, education may require a vast overhaul where leashes are removed, and children are encouraged to explore beyond screens and to spend time in the natural world.

Improved understanding of natural and cultural history can help to build resilience and adaptive capacity crucial to meet upcoming social and environmental upheavals. Knowledge of wild trees, foods, and medicines helps students to feel a link to the land and to respect those who came before us. The lesson of the Iroquois Nation, of forsaking their weapons for peace, and of the colonists, of uniting around a tree which stood for independence and democratic principles, can help to inspire a new generation. Time among trees can also help youth to contemplate the portrayal of success and prosperity many have been raised with, which may be in opposition to a habitable future.

Research on two ends of the spectrum – with Amazonian hunters who chase down prey along rivers – and with American youth who purchase fast foods along highways – has highlighted the fact that trees sustain and heal a wide assortment of individuals on manifold levels. In the Amazon, local recognition of the immense nutritional and cultural value of fruit and game attracting trees leads to protection of fruit trees and forests, contributing to improved livelihoods (Shanley and Gaia 2002). In urban and suburban New Jersey, immersion of youth in woods for tree climbing, trail blazing, invasive species removal, woodland lore, wild edible gathering, and maple sugaring, serves as an equalizer. Over time, seasons spent in woodlands can transform kids from being vulnerable, unruly, or forlorn, to becoming respectful, grounded youth who appreciate what trees and nature offer.

To bridge the gap between people and trees and to restore a connection to nature that is under threat, experience across ecological and social spectrums has highlighted the effectiveness of a pragmatic, place-based approach (Shanley and Medina 2006). Experience with trees you can eat, drink, climb, those which heal, stink, sting, and offer toys, masks, tea, fruit, resins, seeds, and fishing poles – is relevant to everyone's life. Personal connections to trees happen when we touch, taste, smell, see, plant, harvest, and listen to them. For centuries, all cultures have inculcated vibrant traditions, seasonal lore, and legends, and offered recipes and rituals surrounding trees (Schroeder 1992). In post-modern societies, many such intergenerational rituals have receded. Resurrecting the care, respect, knowledge, and skills which early inhabitants displayed toward trees, can help heal the earth and ourselves.

The white pine riots game: reenactment awakens students to history and ecology

The fascinating story of the white pine riots offers a striking opportunity to awaken students to history, ecology, governance, and Indigenous lifeways. To this end, we have developed the White Pine Riots Game based on a fictionalized account of a family making a living off of felling white pines for ship building. As part of this game, students form teams and learn to identify what they would need to construct a ship – white pine for masts; white and red oak for hulls; black locust for nails; and ash or hickory for tool handles. In addition, they need to gather provisions: hickory, walnuts, acorns, greens, forest medicines, and fruits. If students do not pay close attention during the introduction or story, they may not learn how to distinguish one species from another and their team may "expire" during the field-based simulation.

To successfully survive, students need to accurately identify and secure as many of the provisions and ship-building materials as possible. In addition, artifacts such as the white pine flag, pine resin, bones from game animals, and quotes in old bottles from the historic period are hidden in the forest. The finale of the game occurs when "Swamp Law" is called and the colonists take over and drive the King's men out of their forests, claiming the white pines as their own. Jubilee ensues. Each group then shares their findings, discerns which of the gathered species are accurate or not, and discusses the strategies they used and what each artifact signifies. Today, the appropriation of the White Pine Flag by insurrectionists would be discussed, to decipher how symbols of nature can become politically charged, and to help students interpret the evolution of the meaning of the flag over time. In closing, white pine needle tea is served.

What can tree stories, outdoor tree games, and simulations accomplish? First, children have the experience of exploring freely on their own. Using all their senses, they learn to distinguish the difference among pine, spruce, oak, maple, hickory, and locust trees. Before participating, most have not looked closely at the leaves or needles of a tree to distinguish one species from the next, and few have ingested a tree's tea, tapped it for sap, or created a tool from its branches. In the case of the white pine, they also learn the importance of natural history to the founding of a nation. Furthermore, they are introduced to "reading the landscape" and deciphering the meaning of trees, their habitats, architecture, foodstuffs, cultural and Indigenous values, and uses. Mostly they have fun with each other among trees and go away with new eyes. As one student penned in a beautifully illustrated thank you note after participating in the white pine riots game, "I like this different way of learning."

Tree stories, culture, and nature are ties which bind. Most tree, plant, and wildlife species offer significant insights into history, ecology, and culture. In present day society, urbanized populations tend to see only a green blur. Revival of historic and present-day cultural traditions linked to trees is an effective, celebratory way to tap into themes which resonate with all – food, beverages, crafts, medicine, art, dance, song, poetry, theater, and in the case of the white pine, war and peace. Tapping into historical and Indigenous uses, lore, and legend can strengthen the connection between a range of populations with trees – from hunter-gatherers living off of forest goods, to urban citizens and withered, technology-saturated youth.

Conclusion – the legacy of the white pine

From close to 1,000 years ago to the present, the white pine has deeply influenced two daring experiments in democracy and has been central to the building of two nations: the Iroquois Nation and the United States of America. The white pine was the symbol which united warring Native American nations and brought them to a peaceful accord, lasting centuries. Their inspired, sophisticated system of governance subsequently gave rise to key principles embedded in the US Constitution.

Notably, Indigenous nations also held two central principles at the core of their everyday actions which were not transferred to the fledgling democracy – reverence for nature and respect for women. The tree that stood for peace and a lasting accord between tribes was decimated by colonists to support the war effort during the American Revolution and then to build the new nation. Little regard was paid to Native Americans or their respect for and reciprocity with nature or women. In the Iroquois governance system, women are central to decision-making, whereas in the United States, they have been actively marginalized, gaining voting rights only in 1920, close to a century and a half after the Constitution was written.

In the United States, the lack of respect for nature endures. Today, white pine trees are most frequently encountered in monoculture plantations, a faint echo of what they once were. In remote areas, a token giant pine tree may remain, a rare vestige of the ancient grandeur of this venerable species. One white pine found in Maine is listed in the National Register of Big Trees. Its height of 147 feet is equivalent to that of a 15-story high building, and its girth of five and a half feet across matches the height of a fully grown human.

Although the trees of today are a shadow of their bygone past, the white pine continues to prevail as a potent symbol of liberty and freedom from tyranny. Recently, however, it has also been appropriated by right-wing groups challenging democratic elections held in the US. Improved understanding of natural and cultural history can help illuminate what the white pine symbol originally meant for the Iroquois Nation and subsequently what it signified for the revolutionary forces. Misunderstanding and lack of ability to "read" the natural landscape can exacerbate and blur interpretation, particularly regarding reading of the political landscape.

Over the course of the last 250 years, the continued razing of forests and the lack of recognition of the overall value of trees to the health of a nation and its people has had local, national, and global consequences. Extreme floods, fires, drought, and weather events are now commonplace. An estimated 1 million species are now at risk of extinction (IPBES 2019). As the rupture between people and nature has expanded, what has eroded is not only vital knowledge about trees, but the traditions surrounding them – their ceremonies, legends, lore, festivals, and stories. Such traditions enhance understanding and empathy for all living things – attributes which are integral to wise stewardship of trees, and gravely needed at present to support the Earth's faltering life support system.

References

Abrams, M. D., and Nowacki, G. J. 2008. Native Americans as Active and Passive Promoters of Mast and Fruit Trees in the Eastern USA. *The Holocene.* https://doi.org/10.1177/0959683608095581

https://journals.sagepub.com/doi/10.1177/0959683608095581.

Boylan, J. F. 2020. Two Flags Over Maine (and America). *The New York Times*, June 10, p. A27.

Denevan, W. M. 1992. The Pristine Myth: The Landscape of the Americas in 1492. *Annals of the Association of American Geographers* 82(3): 369–385.

Hansen, T. 2018. How the Iroquois Great Law of Peace Shaped US Democracy. *Nature Voices, Native America, PBS.* Available at www.pbs.org/native-america/blogs/native-voices/how-the-iroquois-great-law-of-peace-shaped-us-democracy/#2. Accessed July 20.

IPBES. 2019. *Intergovernmental Science-Policy Platform on Biodiversity and Ecosystem Services.* Summary for Policy Makers: Unprecedented Species Extinction Rates. United Nations.

Jhaveri, I. 2021. The Pine Tree Flag: How One Symbol at the Capitol Riot Connects Far Right Extremism to Christianity. *The Tow Center*, February 24. Available at https://towcenter.medium.com/the-pine-tree-flag-how-one-symbol-at-the-capitol-riot-connects-far-right-extremism-to-christianity-f02314a5f759. Accessed on July 12, 2021.

Klepeis, N. E., W. C. Nelson, W. R. Ott, J. P. Johnson, A. M. Tsang, P. Switzer, J. V. Behar, S. C. Hern, and N. H. Englemann. 2001. *The National Human Activity Pattern Survey (NHAPS): A Resource for Assessing Exposure to Environmental Pollutants.* The Lawrence Berkeley National Laboratory, Berkeley, CA.

Mann, B. A., and J. L. Fields. 1997. A Sign in the Sky: Dating the League of the Haudenosaunee. *American Indian Culture and Research Journal* 21(2): 105–163.

Manning, S. F. 1979. *New England Masts and the King's Broad Arrow.* Wooden Boat Books, Brooklin, Maine.

Peattie, D. C. 1950. *A Natural History of Trees of Eastern and Central North America.* Houghton Mifflin, Boston, MA.

Reilly, M. 2020. The Haudenosaunee Matriarchy and their Influence on Women's Suffrage. *Represent Women*. Accessed on July 30, 2021.

Rutkow, E. 2012. *An American Canopy: Trees, Forests, and the Making of a Nation*. Scribner, New York.

Schroeder, H. W. 1992. The Tree of Peace: Symbolic and Spiritual Value of the White Pine. In *Proceedings of the White Pine Symposium*, pp. 73–83. September 16–18. Minnesota Extension Service, University of Minnesota, Duluth, MN.

Shanley, P., and G. Gaia. 2002. Equitable Ecology: Collaborative Learning for Local Benefit in Eastern Amazonia. *Agricultural Systems* 73: 83–97.

Shanley, P., and G. Medina. 2006. *Connecting through Culture: Transforming Forest Research and Extension for Rural and Urban Relevance in the Brazilian Amazon*. IUFRO, Working Party S6.06–03.

United States Congress. October 1988. *Resolution 331*.

Vietze, A. 2018. *White Pine: American History and the Tree that Made a Nation*. Globe Pequot, Guilford, CT.

Wallace, P. A. W. 1946. *The White Roots of Peace*. University of Pennsylvania Press, Philadelphia, PA.

Whitehead, A. 2021. in Jenkins, J. Capitol violence brewed from nationalism, conspiracies, and Jesus. *The Religious Century, Christian News Service*. Available at https://www.christiancentury.org/article/news/capitol-violence-brewed-nationalism-conspiracies-and-jesus

William, G. W. 2005. *References on the American Indian Use of Fire in Ecosystems*, pp. 1–130. USDA Forest Service, Washington DC.

Winston, K. 2021. The History Behind the Christian Flags Spotted at the Pro-Trump U. S. Capitol Coup. *Religion Unplugged*. Available at https://religionunplugged.com/news/2021/1/6/some-history-behind-the-christian-flags-at-the-pro-trump-capitol-coup. Accessed on July 12, 2021.

4 Bay nuts and pepperwood leaves

Ethically expanding commercial use of a cultural keystone species

Frederica Bowcutt

Introduction

Pepperwood (*Umbellularia californica*) is a hardwood tree native to California and southwestern Oregon with many uses. Makers of bowls, cabinets, furniture, guitars, ukuleles, and hardwood flooring and paneling prize its beautiful wood which is frequently marketed as rare myrtlewood or Oregon myrtle. The leaves, often called California bay or bay laurel leaves, offer medicinal properties and culinary uses. Roasted bay nuts made from its seeds taste like a blend of cacao nibs and dark roast coffee. Culturally important trees like the pepperwood play critical roles, particularly in Indigenous Peoples' cultural identities and survival. However, despite its utility to people, this avocado relative remains commercially underutilized.

This chapter examines past attempts to capitalize on pepperwood leaves and why those attempts have failed or been marginally successful. It also considers contemporary Indigenous efforts to sustainably and respectfully commercialize bay nuts. Using a transdisciplinary approach, evidence is drawn from unpublished notes, images, journal articles, and secondary sources that often fit neatly into one of the following categories: anthropology, botany, forestry, history of science, medicinal botany, phytochemistry, political economy, or Western history. By tracing the earliest attempts to classify pepperwood by colonial botanists to today's Indigenous tribes attempting to commercialize its nuts, we gain a fuller understanding of competing notions about the purpose of economic development and the need for food justice. Indigenous stewardship motivated by a commitment to continued traditional practices could play a significant role in pepperwood conservation in the future.

European bioprospecting

Pepperwood as a subject of Western scientific inquiry has a history rooted in settler-colonialism, the legacy of which continues to shape relations with this culturally important tree. "Although colonialism has [technically] ended, the power relations, systems of meaning, and socioeconomic inequalities that it birthed stubbornly endure" (Go 2016: 185). Aided by the rise of modern

DOI: 10.4324/9780429320897-5

science beginning in the 16th century, various European nation-states embraced "a new ethic sanctioning the exploitation of nature" (Merchant 1980: 164). This set the stage for "a progressively rationalized 'cost-profit calculus'" (Moore 2018: 244) in the service of empire building that depended on "mapping and cataloguing the globe as a potential storehouse of wealth" (Moore 2017: 605). Botanists aided European colonial expansion as seekers of "green gold" in the form of new crops that could provide "renewable profits long after gold and silver ran out" (Schiebinger 2004: 7).

Transnational merchant capitalists as well as nation states benefited from global bioprospecting efforts (Moore 2018). The work engaged both field botanists and sedentary men working methodically in collections of dried plant specimens. European herbaria accumulated these specimens over hundreds of years from multiple field botanists who, through their contributions, enabled the development of centralized knowledge. Their universalized system of naming and classifying accelerated the study of global plant diversity, making it easier for botanists to identify plants reliably. "[C]olonial profits depended largely on natural historical exploration and the precise identification and effective cultivation of profitable plants" (Schiebinger and Swan 2005: 2). Herbaria, along with botanic gardens, developed as key research tools for colonial botany as a branch of Western science.

Colonial botany, from its inception, involved the search for useful plants that could be profitably commercialized. Bioprospectors or seekers of valuable biological resources were often included on important early voyages like the Vancouver Expedition lead by Captain George Vancouver (1757–1798) of the Royal British Navy. The primary goals of the voyage were cartographic, but the naturalist onboard, Archibald Menzies (1754–1842), collected herbarium specimens of many species new to Europeans. Sometime between 1792 and 1794, Menzies collected the first herbarium specimen of pepperwood somewhere along the California coast between Trinidad Head in Humboldt County and Santa Barbara (Eastwood 1939, 1945).[1] Charles Sprague Sargent, author of *The Silva of North America*, claimed Menzies likely saw it first in November 1792 along the San Francisco shoreline (Sargent 1895). Menzies introduced Europeans to new "vegetable treasures" through his specimens and piqued interest in their commercial potential (Nuttall 1842: 88). He was well qualified for bioprospecting, having served as the naturalist on multiple other British voyages on the recommendation of Joseph Banks (1743–1820) who advised the King of England on botanical explorations.

Often remotely, botanists participated in the inventorying of western North American botanical resources during a time of unresolved land claims between Britain and its former colony. Many of the first pressed specimens of pepperwood were sent to the herbarium at Kew Gardens, a London-based research facility, which became essential to the establishment of 19th century plantation economies in various British colonies (Brockway 2002). William Jackson Hooker and G. A. Walker Arnott, working as botanists there, served the Secretary of State for Colonial Affairs. They wrote the original scientific description of pepperwood based on Menzies' herbarium specimen (Eastwood 1945). In 1833, Hooker and Arnott tentatively placed it in the genus *Tetranthera*, but the specimen they were working from lacked mature flowers, which were needed to make a more definitive determination (Hooker and Arnott 1840). Without sufficient understanding of the reproductive organs, its relationship to other plants perplexed early botanists, resulting in a proliferation of common and scientific names (Table 4.1). Multiple attempts were made to put pepperwood in various tropical genera based on shared foliage characteristics. After many name changes, botanists agreed on the current name of *Umbellularia californica*, which was first introduced in 1842. It was classified in the primarily tropical and subtropical laurel family (Lauraceae), which includes many economically valuable plants such as avocado, camphor, cinnamon, and sassafras (Jepson 1909; Judd et al. 2016). Botanists engaged in a tradition of imperial plant naming through universalizing methods designed to

Table 4.1 English common and scientific names of Pepperwood

English common names	*Scientific Names followed by authors of the name and then (year first introduced)*[a]
balm of heaven	*Laurus pauciflora* Blume (misapplied in 1826)
bay, bay laurel, bay tree, bay-tree, baytree	*Tetranthera californica* Hook. & Arn. (1833)
black myrtle	*Oreodaphne californica* (Hook. & Arn.) Nees (1836)
cajeput tree	*Laurus regia* Douglas (1836)
California bay, California-bay, California bay laurel, California laurel, California-laurel, California olive	*Drimophyllum pauciflorum* Nutt. (1842)
cinnamon bush	*Umbellularia californica* (Hook. & Arn.) Nutt. (1842; **currently accepted**)
Coos Bay laurel	*Litsea californica* (Hook. & Arn.) Benth. & Hook.f. ex B.D. Jacks (1894)
headache tree	**other synonyms no longer used:**
laurel, mountain laurel, Oregon laurel	*Laurus californica*[b]
myrtle, myrtlewood, myrtle-wood, myrtletree, Oregon myrtle, Oregon-myrtle, Pacific myrtle	*Laurus regalis* Hort.
peppernut tree, pepperwood	*Laurel silvestre*[c]
sassafras laurel, Californian sassafras	*Persea causticans* Pasq.
spice bush, spice-bush, spice-tree	
white myrtle	
yellow myrtle	

Sources: Abrams 1910; Abrams and Ferris 1944; Bois 1919; Clarke 1985; Hedrick 1919; Howard 1992; Jepson 1910; Nassini et al. 2012; Sargent 1922; Snow 1903; Stein 1990; Sudworth 1908; United States Senate and United State War Department 1856; Van Tieghem 1892; Wright 1902. Specific sources for specific names listed available on request of the author at bowcuttf@evergreen.edu.

a For a partial history of synonymy, see Jepson (1910: 242).

b Used as a synonym without an author attribution in Stillman (1880: 3749). Now used for an extinct plant known from the fossil record.

c The name used by Spanish Californians in the 1700s (Jepson 1910: 245).

objectify nature through classification based on shared characters and differences (Foster 2017). Naming also served as a way of claiming botanical resources for exploitation (Foster 2017; Schiebinger 2004).

Botanical illustration emerged in the early modern era as "a central practice for investigating colonial nature and incorporating it into European science. . . . Seeing was intimately connected to both knowing and owning. Images of plants . . . were instruments of possession" (Bleichmar 2006: 82). They also "reinforce objectivity" and transform plants into objects worthy of scientific study (Foster 2017: 36). Building on earlier traditions, Charles Edward Faxon meticulously captured the diagnostic characters of pepperwood in his illustration published in Charles Sprague Sargent's 1895 book on the trees of California (Figure 4.1).[2]

Figure 4.1 Line drawing of *Umbellularia californica* (Hook. & Arn.) Nutt. with reproductive organs by C. E. Faxon. Detail number 5 shows a longitudinal section of a flower and detail number 12 shows a fruit with the nut inside (Sargent 1895).

Source: From Sargent, *The Silva of North America*, vol. 7, Plate 306 (1895).

The typical inflorescence type pictured in the upper left is called an umbel and inspired the genus name, *Umbellularia*, which is derived from the Latin word for sunshade (*umbella*) (Pratt 1922). The bisexual flowers in these small clusters in the axils of the alternating upper leaves appear from December through May (Munz 1973).

Latin as the universal language of science and the initial basis of botanical nomenclature excluded women, people of color, and other marginalized people with limited access to formal education (Schiebinger 2004; Stearns 2004). Widespread interest in plant taxonomy among middle- and upper-class white women sparked fear of the feminization of botany in the 18th and 19th centuries. In an article entitled "Is Botany a Suitable Study for Young Men?" published in the journal *Science* in 1887, the author sought to attract men to the field claiming "botany ought to be ranked as one of the most useful and most manly of studies" because it requires mental discipline, physical fitness, a "robust character", and a desire to be useful through the cultivation of medicinal plant knowledge (Adams 1887: 116–117).

Early investigations of Indigenous foods

Several decades after Menzies' botanical studies of North America, the London Horticultural Society sent David Douglas (1799–1834) as a resource scout to seek out valuable food and garden plants in North America. Douglas remarked in October 1826 that pepperwood was a "fine tree" with medicinal and horticultural potential and proposed making it "an inmate of English gardens" (Nisbet 2009: 124).[3] He first encountered pepperwood in the Umpqua River watershed on his way to northern California. Douglas was fed its stimulating roasted nuts by his Indigenous guides after an injury and becoming fatigued from heavy exertion (Douglas 1914). The physical demands bioprospectors faced and overcame on their travels were widely reported to lend their work credibility. Many were celebrated as heroes and those who lost their lives in the process, like David Douglas, were often seen as martyrs for science and the empire (Schiebinger 2004). However, among Indigenous People, such men were often recognized as threats that could contribute to their eventual dispossession of land and the usurpation of valuable plants. Primitive accumulation occurred in part through genocide. Efforts to exterminate California Indians escalated in the mid-19th century, particularly after the discovery of gold (Aron 2015).

Based largely on ethnographic research with survivors, American anthropologists during the late 19th and early 20th centuries recorded many traditional uses of pepperwood. In 1902, an ethnographer who studied Indigenous plant uses in Mendocino County wrote "some of the native food stuffs, such as the bay nut . . . might be introduced into our markets or at least into our home gardens for human consumption" (Chesnut 1974: 297–298). The flowers appear to be primarily pollinated by small insects (Stein 1990). As the fruit matures, it turns from green to dark purple. Ripening typically occurs in October and

November. Trees can be intermittent seed producers, but in the right growing conditions they can be "heavily laden with fruit" (Sudworth 1908: 331). The avocado-like fruit is attached by a somewhat enlarged short stalk (Figure 4.1). A thin layer of flesh covers a single, large, smooth pit inside. In good years, well-formed trees, particularly in full sun, bear bumper crops.

Multiple tribes in northern California have a long history of using its nuts as a food source, including the Karuk, Kashaya Pomo, Ohlone,[4] Potter Valley Pomo, Tolowa, Yuki, and Yurok (Moerman 2010). The nutritious value of the nuts apparently is similar to walnuts (Clark 2015). Whole bay nuts were, and still are, stored in large quantities for later use after removal of the outer fruit flesh. When not prepared properly, bay nuts are toxic. Methods to address this fact were developed and practiced by Indigenous women, and in the process, they made an otherwise inedible nut delicious and nutritious. The nut is roasted, cracked open as needed, and then pounded into a flour, which is then cooked or not. The Karuk and other tribes would and still do sometimes eat the nuts after roasting but before pounding, often by themselves or with oak acorns (Schenck and Gifford 1952). Although the green flesh attached to the nut can irritate the mouth, once the fruit ripens and turns purple, the fleshy covering is edible raw.[5] The Kashaya Pomo ate it "after it had been loosened and split while the nuts lay drying in the sun" (Anderson 2005: 277). The Miwok in Yosemite Valley constructed granaries to cache bay nuts, as did many other tribes elsewhere (Anderson 2005). Drawing from granaries during winter, the Potter Valley Pomo used the nuts to diversify a diet emphasizing acorns (Welch 2013). They also used them to flavor the bland meal made from California buckeye (*Aesculus californica*) nuts (Welch 2013).

Salvage anthropologists like John Hudson sought to record the pre-contact cultural practices of Indigenous Peoples "before they vanished or became too assimilated" (Baldy 2018: 5). Beginning in 1890, he maintained extensive ethnobotanical notes based on nearly 50 years of personal relations with Potter Valley Pomo consultants in northern California.[6] In his unpublished notes, he documented their name for the pepperwood tree as *be-hi ka-li*, however the more standardized phonetic spelling is *behem xale* (Hudson n.d.a; see Table 4.2).[7] Roasted as needed in hot embers, the nuts were then cracked open, peeled, and pounded with a stone pestle while still warm. Once pounded into the consistency of brownie dough, it was and still is rolled or otherwise formed into bite-sized pieces called *behechni*, which hardens as it cools (S. Smith-Ferri pers. comm. 2018). Alternatively, the dough may be cooked after the roasted nuts are processed (Hudson n.d.b). Hudson explained that the "cakes made of the laurel nut meal" are of "flinty hardness" and "useful to messengers and travelers on the trail" as a nourishing endurance food (in the contemporary vernacular, power bar) (Hudson n.d.c: 80). With salt as a preservative, the oily meal could be made into unleavened "laurel nut bread" (Hudson n.d.d: s.n.). Hudson claimed it "keeps indefinitely" and could be reheated or moistened to make it easier to chew (Hudson n.d.c: 80). It was eaten in small to moderate quantities "as a stimulant or condiment" (Jepson 1910).

Table 4.2 Some of the many Indigenous names associated with Pepperwood

behem? (Kashaya Pomo)
behkas' or for the nut beh-e (Southern Pomo)
bihem xale for the tree (Northern Pomo)
bok (Numlaki)
sícils for the nuts (Hanis and Milluk)
mopák (Salinan)
pahip or for the nut *pah* (Karuk)
psha'n, psha'an (Chumash)
so-e-bä (Concow)
waegaent for the tree (Hanis)

Sources: Chesnut 1974; Goodrich et al. 1980; Immel 2007; Northern Pomo Online Talking Dictionary 2020; Phillips 2016; Schenck and Gifford 1952; Smith 2014; Timbrook 1990.

Mirroring earlier taxonomic treatments of humans, Hudson placed Indigenous Peoples below "civilized man". Justification for their displacement during European colonization and American expansion rested on a foundation of racial ideologies (Bauer 2016). Swedish taxonomist Carl Linnaeus in the last edition of *Systema Naturae* published in 1766, classified humans as distinct varieties with Native Americans characterized as choleric people prone to being righteous, stubborn, zealous, and regulated by customs, while people of European ancestry, in contrast, were categorized as sanguine, gentle, inventive, and governed by laws (Linné 1766). Linnaeus believed science to be a savior that would ultimately help Scandinavia and other famine-stricken areas through the discovery of new food plants. He instructed his students who traveled outside of Sweden "to study local peoples' knowledges of the natural world and their manufacturing techniques" (Koerner 1996: 152).

Although bay nuts were not among them, a variety of other Indigenous food plants from the Americas adopted after 1492 in Africa, Asia, and Europe contributed to radical demographic changes globally as part of what Alfred Crosby called the "Columbian exchange", by improving the quantity and quality of foods available (Crosby 2003; Weatherford 2010). Some novel "New World" food plants like corn and potatoes contributed to significant geographical shifts in political and economic power, generally from southern regions like Spain and Italy to northern regions including Germany, Britain, and Russia. Other North American foods like tomatoes and chilis remade the cuisine of China, India, and the Mediterranean region, among other areas. Many became critically important to an industrialized food system and previously overlooked Indigenous foods from the Americas continue to be introduced into global trade, such as amaranthus (in the 1970s) and quinoa (in 1986) (Weatherford 2010). With the benefit of hindsight, "native crops of America have produced a far greater impact than the mountain of silver" in the Andes plundered by the Spanish (Weatherford 2010: 84).

The study of Indigenous food culture offered salvage anthropologists a way to study human social evolution as a universalizing theory of progress which "assured ultimate improvement" (Dippie 1982: 105). The perception that Native people are static primitives frozen in time persisted as a defense of a dynamic modernity that transformed food and medicine into commodities as part of progressive ideology.

Like other late 19th and early 20th century salvage anthropologists, John Hudson considered Native Americans as incapable of adapting to a society that objectifies plants and seeks to manipulate and commodify them for financial gain. Hudson saw them as passive gleaners, failing to see how Indigenous Peoples skillfully manipulated plant populations with burning, pruning, and digging to favor those species they used.[8] "The Pomo was part of nature, never master of it, he cultivated nothing. Like the birds he ate what grew as it grew. . . . It was for civilization to develop big potatoes, big onions, beans and other time savers" (Hudson n.d.d: s.n.). Hudson claimed "[t]he White man alone is supreme. The Indian believes himself a part of a great consciousness shared by every living thing" (Hudson n.d.d: s.n.). Typical of Euro-American intellectuals at the time, Hudson viewed traditional Indigenous cultures as static relicts comprised of superstitious peoples vanishing with the inevitable and rapid U.S. industrialization and urbanization of the 19th and early 20th centuries.

According to two historians of science, "if a single vision predominates in modern Western society, it is that of a passive and disempowered nature, slave and victim of human agency" (Jardine and Spary 1996: 4). But that view is not held universally. According to traditional Tribal stories, the Creator placed plants like pepperwood on Earth for the original people to use respectfully (Bauer 2016). Women primarily harvested plant-based foods and developed and practiced the culinary technologies needed to consume them. And they "used the fruits of their labor to assert a political role in their communities" (Bauer 2016: 19). Indigenous economies were built on a gendered division of labor (Bauer 2016). However,

> anthropologists, archaeologists and historians have sometimes been slow to focus on and acknowledge the essential contributions of women in food production. Likewise, they have tended to overlook the complexity and sophistication of the knowledge that women have held and applied as food producers, herbal medicine specialists and resource managers.
>
> (Turner 2003: 134)

Indigenous medicinal knowledge and patent rights

Ironically, given the racial hierarchies and heteropatriarchy undergirding colonialism, bioprospectors often drew from ethnographic research on plant uses based on the premise that "plant-based medicine was more highly developed among the Indians of the New World than it was in Europe" (Brockway 2002: 109). In the 16th century, Spaniards were well aware that they "depended on the authority and knowledge of the Indigenous people" they colonized in the Americas (Barrera 2002: 167). In addition to their utility in empire building and treating colonists, new Indigenous plant medicines held the promise of commercial opportunity (Jardine 1999). Indeed "a league of Andean municipalities" attempted to block British bioprospectors from appropriating cinchona through a national law passed in the 1850s prohibiting exportation of plants from Peru, Bolivia, and Ecuador without a license (Mushita and Thompson 2007: 26–27). Also called fever tree, its bark is a source of quinine, which is effective in treating malaria. European colonization would have likely failed in many areas, including India, without cinchona plantations (Brockway 2002).

For millennia, Indigenous Peoples have used pepperwood leaves to treat a variety of medical conditions, including earaches, head lice, menstrual cramps (Adams and Garcia 2006; Strike 1994), nasal and chest congestion, meningitis (Gould 1900), nervous conditions, pain associated with the afterbirth, rheumatism, skin conditions, stomachaches, toothaches, and wounds (Moerman 1998). Its most commonly recorded use is as an analgesic for headaches, including by the Cahuilla, Miwok, Kashaya Pomo, and Yuki (Moerman 1998). The leaves were used in infusions, decoctions, poultices, or simply crushed and inserted into a nostril or placed under a hat (Chesnut 1974; Moerman 1998).[9] Multiple tribes have used a decoction of the leaves to treat colds and/or sore throats, including the Chumash, Karuk, Kashaya Pomo, and Pomo (Carranza et al. 2015; Moerman 1998; Timbrook 1990). Leaves used in steam baths and as a poultice also have antimicrobial properties based on the ethnographic record (Moerman 1998). Some tribes considered pepperwood a panacea (Moerman 1998; Schenck and Gifford 1952). The Yuki burned the leaves "to take bad luck away" (Moerman 1998: 579). The Karuk use them to drive evil spirits away during the Brush Dances by placing the leaves on the fire (Baker 1981). The leaves are still used for protection against harmful forces. Some tribes use a burning pepperwood branch for spiritual cleansing during ceremonies (Salmón 2020). Due to its ceremonial uses, Kashaya Pomo still consider pepperwood "very sacred" (Goodrich et al. 1980: 91).

Indigenous knowledge of plants often became obscured in the enterprise of a universalizing botany as local, place-based knowledge became marginalized and/or coopted (Schiebinger 2004). During Spanish colonization, the Franciscan Fathers at Mission San Antonio de Padua in present day Monterey County California adopted pepperwood leaves into their pharmacopeia (Garriga 2001; Heinsen 1987). They improvised on Salinan medicinal practices by combining the leaves with nutmeg, cinnamon, and olive oil to make a heated poultice used to help people suffering from fits to regain consciousness (Garriga 2001; Immel 2018). Later, Euroamerican settlers concocted a salve for rheumatism from pepperwood leaf-infused lard, emulating its use by Indigenous Peoples newly experiencing this inflammatory condition provoked likely in part by significant dietary changes resulting from white settlement (Chesnut 1974).

Historically, colonists often gained information on uses through their Native wives (Barrera 2002), a fact that was rarely acknowledged once plant descriptions were incorporated into botanical treatises. This was a recapitulation of European history whereby the contributions of community-based folk healers, often female, went unacknowledged or dismissed. In his German herbal *De Historia Stirpium*, Leonhart Fuchs (1502–1566) used accurate botanical illustrations based on direct observation to aid correct identification of medicinal plants and to assert the superiority of his knowledge. While appropriating traditional European botanical knowledge in his written descriptions, he used his class privilege to discredit women and other illiterate people claiming that objectivity and rational thought was beyond their reach. He dismissively referred to female herbalists as "foolish and superstitious old women who gather herbs and roots" (Egerton 1983: 276–277).[10] Yet, he simultaneously profited personally from their collective herbal knowledge, becoming a successful physician skilled at treating the plague in his affluent patients. Gender-based ideologies empowered men like Fuchs to dismiss women's "vision, skill, and hard work" over the millennia to support a mythology about men as the only people involved in the skillful procurement of healing herbs (Johnson 2014: 134–135). A concerted effort had been made in Europe during the 18th century and earlier to expunge medicine of "popular healers, many of them women" known as "lady doctors" who learned how to use medicinal plants through oral traditions (Eamon 1994: 358).

According to contemporary Native American scholar Cutcha Risling Baldy, anthropologists purposefully made female doctors nearly invisible because their existence and even dominance in some Indigenous communities challenged academic universalizing that normalized heteropatriarchy (2018). They "often oversimplified, ignored, or actively erased from the anthropological record" pervasive gender equality and instead created "a mythological Native past that mirrored the heteropatriarchal structure of settler colonial society" (Baldy 2018: 80, 30). Indigenous cultures challenged the narrowly defined male/female binary of Victorians that viewed men as uniquely capable of being wise and competent (Baldy 2018). Despite the prevalence of Indigenous women in leadership roles, including as healers, anthropologists emphasized the importance of medicine men (Baldy 2018). John Hudson's wife Grace Hudson, who was a renowned painter, drew on multiple romantic tropes by portraying "John, Scott's Father" as a noble savage with a California bay laurel wreath on his head that harkens back to head wreaths used by ancient Grecians and Romans made from the Mediterranean laurel (*Laurus nobilis*). Many Indigenous healers in northern California used and still use bay laurel leaves for spiritual purification and protection.

Figure 4.2 "John, Scott's Father," also known as Kol-ba, a well-known healer from Pinoleville wearing a head wreath of pepperwood leaves, in a 1900 portrait by Grace Hudson.

Source: Courtesy of the Collection of the Grace Hudson Museum & Sun House, City of Ukiah, California.

Pepperwood leaves contain a variety of chemical compounds that can act on the human body, including alkaloids, anthraquinones, flavonoids, and terpenoids (Carranza et al. 2015). The chemical compounds listed and others possess antimicrobial properties (Salmón 2020). Based on studies on mice, rats, and fish, another compound in the leaves called umbellulone has toxic qualities (MacGregor et al. 1974). In 1842, Thomas Nuttall noted that "the foliage gives out, when bruised, a most powerful camphorated odour, which from its pungency is capable of exciting sneezing" (Nuttall 1842: 88). The leaves, wood, and even sawdust can cause headaches for some sensitive individuals due to their potent smell (Benemei et al. 2009; Stillman 1880; R. Saecker pers. comm. 2020). By 1857, a New York doctor claimed a cordial made from California laurel could cure treatment-resistant nervous diseases and "restore vigor, animation, and the perfection of good health to a constitution shattered beyond the hope of recovery" (Monnett 1857: 35). One man who had been "at death's door" miraculously recovered and at nearly 80 years old, he reportedly enjoyed the vitality of a 30-year-old. He claimed to have used the "elixir of life" ever since his initial treatment to maintain his good health based on knowledge traced back to an Indigenous woman (Monnett 1857: 118–119). The New York doctor also claimed his tincture addressed the warning symptoms of tuberculosis, including a hacking cough and bronchitis.[11] He imported the leaves directly from a supplier in San Francisco because normally, they could "not to be obtained without great trouble and expense" (Monnett 1857: 35).

A *Year-Book of Pharmacy* from 1875, reported "some of the native Californians have peculiar ideas about this tree" which they believe can cause or aggravate asthma (British Pharmaceutical Conference 1875: 229–230).[12] The account goes on to note successful use of the oil in treating nasal congestion and nervous headaches. In 1879, a J. M. Stillman distilled the oil from the leaves and compared the oil's chemical composition to turpentine and its odor to camphor (Stillman 1880). By 1902, some patented medicines reputedly contained oil primarily derived from pepperwood leaves (Chesnut 1974). The 1903 pharmaceutical products catalog of Frederick Stearns & Co. based in Detroit, Michigan offered a fluid extract made from its leaves which it advertised as "efficacious in nervous headaches, intestinal colic, and atonic diarrhea", under the common name California laurel (Frederick Stearns & Co. 1903: 215).

In 1885, Florence Saltonstall reported on her experimentation with using pepperwood homeopathically (Saltonstall 1885). Using an older scientific name for the tree, a remedy called *Oreodaphne californica* appeared in Dr. William Boericke's manual on homeopathic remedies. First published in 1901, the widely used reference text listed it as a treatment in a low potency form for "neuralgic headache, cerebro-spinal meningitis, atonic diarrhoea, and intestinal colic" (Boericke 1906: 486–487). In an address to the alumni of Hahnemann College of the Pacific in 1911, the San Francisco homeopath advocated for using native plants in medical practice because they "often are better adapted to local disease types than other remedies", and as a regional example, he noted *Oreodaphne californica* as effective against headaches (Boericke 1911: 187). Placed-based notions of healing inserted themselves into American life informed by Indigenous cultural practices.

Practitioners of botanical medicine and homeopathy in the U.S. enjoyed widespread popularity during much of the 19th century (Ackerknecht 2016; Libster 2004). However, increasingly, they lost credibility after the formation of the American Medical Association in 1847, with the rise of bacteriology in the late 19th century, and the expansion of modern medicine during the early 20th century (Ackerknecht 2016; Haller 2009). Pepperwood-based medicines failed to gain serious traction in the dominant American pharmacopeia. In contrast, despite widespread condemnation of "Native healers as ineffective and anti-modern", including by the federal Office of Indian Affairs, "traditional healing practices persisted and thrived during the Great Depression" among the Pomo (Bauer 2016: 106, 111). Multiple tribes within the tree's range resisted wholesale adoption of allopathic medicine and continued using bay laurel leaves for healing, and many still do.[13] Despite the legacy of conquest, Indigenous People resisted the relentless efforts to colonize bodies and minds by defending their traditional healing practices. California Indians asserted that their "folk-healing methods . . . often worked better than Settler medicine" (Bauer 2016: 8).

Early conservation efforts

Leading up to and during World War II (WWII), logging increased radically to supply wood for the war effort. Extraction of Douglas-fir (*Pseudotsuga menziesii*) and coast redwood (*Sequoia sempervirens*) from northern California and southwestern Oregon forests also impacted pepperwood, a common associate. *Umbellularia* is a monotypic genus restricted to the west coast of North America, meaning there's only one species in the genus (Spongberg 1975). It grows in the Coast Ranges from southern California to the Umpqua River Valley of Oregon. Its distribution extends east to the foothills of the Cascade Mountains and south along the western slope of the Sierra Nevada growing near sea level up to nearly 5,000 feet elevation (Stein 1990). A tree with "broad ecologic versatility", it can tolerate a wide range of climatic conditions, from high precipitation to drought (Stein 1990: 827). In dry regions, it's often restricted to riparian zones, alluvial soils, north-facing slopes, or spring fed gulches (Stein 1990). Stands along the Eel River in northern California are among the best developed (Jepson 1909). A variety of forest types can support pepperwood (Stein 1990).

Although pepperwood is common within California and southwestern Oregon, botanist Alice Eastwood described a new variety in 1945: *Umbellularia californica* var. *fresnensis* found only in Fresno County in the southern end of the Sierra Nevada (Eastwood 1945). This specimen is distinctly different with "longer and denser" hairs on the inflorescence stalk, but further research is needed (Flora of North America Editorial Committee 1997: 31). According to the International Union for Conservation of Nature and Natural Resources, this variety is vulnerable, resulting in its placement on their Red List (IUCN 2021). Eastwood served as the herbarium curator at the California Academy of Sciences in San Francisco from 1892 to 1949, until her retirement at the age of 90 (Dakin 1954). Considered inappropriately bold given her gender, the largely self-taught Eastwood secured the job with the support of Katherine Brandegee another non-conformist female botanist (Keeney 1992). Eastwood became "one of America's leading botanists and a conservation advocate" (Riley 1999: 43).

Although the exceptional Alice Eastwood proved the rule, professional opportunities remained limited for women throughout much of the 20th century. Some instead became club women to fight for conservation like Lilla Leach who directed Save the Myrtlewood League, Inc. from 1945–1948 (Riley 1999). Leach and others took issue with the impact of increased logging on aesthetically pleasing pepperwood groves. Before the group dissolved in 1964, the organization partnered with timber companies and public agencies to set aside scenic and accessible groves as public lands (Casey 1974). Makers of furniture, instruments, bowls, and various knickknacks still celebrate myrtlewood as a rare and unique commodity and try to use exchange value to make a case for conservation.[14] Although small-scale, wood products from *Umbellularia californica* "provide a . . . significant source of income for small mills and crafters in southwest Oregon and California" (Lang 2018: s.n.).

Postwar economic growth and its human health impacts

Post-WWII suburbanization led to continued logging pressure, and by the 1950s, foresters in the western U.S. began using herbicides to weed out native hardwood species like pepperwood to favor softwood production due to the higher profit margins involved (Bowcutt 2015). Postwar economic growth also accelerated agricultural industrialization, contributing to the rise of the industrial diet which "had the net effect of degrading food from a nutritional standpoint" (Winson 2013: 29). A more simplified diet of adulterated foods contributed to a global health crisis (Winson 2013). Rates of diabetes rose rapidly in the 20th century and continue to rise among Americans, with rates among Indigenous Peoples being particularly high (Coté 2016; Winson 2013). Radical dietary changes in the late 20th century caused diabetes rates to skyrocket among the Karuk people whose traditional foods include bay nuts (Norgaard 2019). Malnutrition also increased in Indigenous communities with greater dependence on processed foods (Coté 2016). Beginning with white settlement, "the loss of access to traditional foods had a devastating impact" on Indigenous Peoples (LaDuke 2005: 194).

The American century that fueled an expansion of industrial agriculture through scientific research and mechanization transitioned into a new era of bioprospecting shaped by neoliberal values (Moore 2018). Overuse and other unskillful uses of antibiotics in medicine resulted in the development of antibiotic resistance beginning in the 20th century, creating new potentially lucrative markets. In 2015, several researchers scientifically validated that some of the traditional, Indigenous medicinal uses of pepperwood had merit. To protect their legal right to patent phytochemicals, they claimed to be "the first to report the antibacterial activity of extracts from [its leaves and bark] . . . and illustrate [its] promising anti MRSA potential" (Carranza et al. 2015: 29). Patent laws can be used to deprive Native American nations from benefiting economically from their plant knowledge. Even in the late 20th century, ethnopharmaceutical information ungirds the U.S. pharmacopeia with its significant dependence on plant products and simulations of them (Lewis 2000). "The patent is a symptom of scientific culture that privileges the fiction of the genius" while making invisible the many contributors to knowledge construction (Osseo-Asare 2014: 203).

Biocultural conservation

> The history of capitalism has been a chronicle of economic expansion interrupted and accompanied by the kinds of malfunctions that derive precisely from the absence of a mechanism capable of bringing the private aims of capital into harmony with the social needs of the community.
>
> (Heilbroner 1980: 132)

That said, "reality is not merely what 'is,' but what we make it . . . to depict things as static or changeless is to disregard or violate the essence of their being" (Heilbroner 1980: 31–32). Multiple California tribes are assessing the viability of revitalizing traditional food plants to address health issues linked to the industrial diet and promote economic opportunity for their members. Research indicates "the prevalence of type 2 diabetes tends to be much lower" among Native people who eat their traditional diet (Winson 2013: 233). Novel Indigenous foods are also trendy among foodies. Chefs of fashionable California restaurants purchase wild foods from professional foragers (Goldstein 2013; Raskin 2012; Sacks 2013). Roasted bay nuts taste like dark roasted coffee and cacao nibs combined, making them highly desirable in niche markets.

However, given entrenched patterns of exploitation in U.S. society, how might tribes commercialize bay nuts to foster Tribal economic development and Indigenous food sovereignty without initiating unintended negative environmental and social impacts? Wild foraging and commercial production of novel Indigenous foods could result in continued cultural appropriation and reduced access for Indigenous People. "As people have been exploring the possibilities for commercialization of nontimber forest products in California, developing and shifting markets for products have sometimes created conflicts among ethnic or cultural groups for access to plant and fungi resources" (Weigand 2002: 83–84; see also Tsing 2015). Industrial production could displace small-scale producers, if successfully adopted to meet market demand. For example, cheaper, industrially produced wild rice from diked paddies in California out competed wild rice traditionally harvested by Indigenous People, which was an important source of income (LaDuke 2005). Global trade can also raise the economic value of traditional foods to the point that they become precious export crops that are too valuable to eat by Indigenous producers, as has been the case with quinoa (McDonell 2018). The world over, increased production of cash crops has historically resulted in less self-sufficiency and more market dependency.

Most of us don't "eat our landscape", but instead depend on the landscapes of others for our food (Salmón 2012: 154). In 2012, the Karuk, Yurok, and Klamath tribes received support from the National Institute of Food and Agriculture to partner with the University of California, Berkeley's herbaria to "test innovative, community-led solutions to enhance food security and food sovereignty" by revitalizing access and use of their traditional foods like bay nuts. The five-year, multimillion-dollar project involved stakeholders "in project development, implementation and evaluation" (USDA n.d.: s.n.). Valued trees enhance conservation efforts by conjoining ecology with cultural thriving. Based on their studies on chestnut trees in Turkey, Jeffrey Wall and his collaborators concluded that "both biological and cultural, or biocultural, diversity is necessary for meaningful human survival" (Wall 2018: 14). Treating plants as kin results in healthier ecosystems according to Indigenous Peoples the world over (Senos et al. 2006). According to the Nuu-chah-nulth, the first peoples of Vancouver Island, "in a seamless and interconnected web of life where all life forms are revered and worthy of mutual respect . . . plants are regarded as . . . kinfolk, not as a commodity that can be exploited" (Coté 2016: s.n.). In an article titled "Indigenize Conservation," Melissa K. Nelson argues "we need stewardship that is embedded in land-based cultural practices known as traditional ecological knowledge (TEK)" (Nelson 2021). Senos et al. claim that integrating TEK into restoration work enhance[s] ecosystem functioning and "the ability to sustain indigenous or local peoples' economies and cultural practices" (2006: 393).

Although pepperwood is far from endangered, the traditional Indigenous knowledge of how to transform the nuts into brownie-like bay nut cakes is becoming less commonly practiced (Nelson 2004). Peppernuts, as they are also known, require a lot of time to gather and process – plus you need access to productive trees. The traditional Indigenous practices of fostering tree health and abundant nut production are also threatened. As a result, "many younger people are not learning this ancient food process, and as elders pass on, these practices pass on as well" (Nelson 2004: 33). Through annual celebrations of bay nuts as "[o]ne of the most delicious and unique California Indian foods," The Cultural Conservancy in San Francisco is working to revitalize those practices (Nelson 2004: 32). However, commodification of the nuts even as a means to revitalize traditions is a sensitive issue because traditional foods are sacred. According to Walter R. Echo-Hawk, "In the cosmology of Native American gatherers, plants hold an honored place as the foundation for human and animal life" (Echo-Hawk 2010: 370). Charlotte Coté, another Indigenous scholar, explains that "Indigenous philosophies are grounded in the understanding that humans and the environment are bound in a relationship of reciprocity, respect, and obligations – not coercion and domination" (2016: s.n.). Some tribes seeking to fulfill their moral responsibilities are resorting to "[W]estern concepts of intellectual and cultural property rights to safeguard their cultural knowledge and practices from exploitation by outsiders" (Nelson 2004: 33). However, establishing intellectual property rights for knowledge held collectively by a multiplicity of tribes is challenging (Osseo-Asare 2014).

In addition, multiple ecological challenges need to be addressed if bay nuts are to become a profitable commodity as hoped by some tribes. For example, various diseases and pests introduced through global trade in nursery plants and wood products threaten pepperwood, including laurel wilt (caused by the Asian fungus *Raffaelea lauricola* introduced to the se U.S.). Introduction rates have increased significantly with reduced regulations imposed by free trade policies. Due to over a century of fire suppression, the fire sensitive pepperwood is more abundant now, which may contribute to disease and pest threats because higher density often facilitates spread (Bowcutt 2015). Other challenges to commercialization include inconsistent fruit production year to year and uneven fruit ripening (a single fruit can be simultaneously rotten and not ripe). Although pepperwood is adapted to the droughty west coast of North America, it is vulnerable to climate change.

Whether folk value can aid in addressing the contemporary ecological crisis in California and southwestern Oregon caused by sudden oak death remains in question. The disease-causing pathogen *Phytophthora ramorum* sporulates on a wide variety of host species. However, some species play more important roles, such as pepperwood, which exhibits the highest potential for spreading the disease (Cobb et al. 2013; Meentemeyer et al. 2004). This leads some plant pathologists to recommend removal to reduce risk to neighboring oaks (Swiecki and Bernhardt 2008). Some affluent home owners in the San Francisco Bay area are killing pepperwood to eliminate disease vectors that pose a threat to coast live oak (*Quercus agrifolia*), which as mature majestic trees can contribute significantly to property value (Bowcutt 2015). This illustrates that "human culture and nonhuman nature can affect one another and . . . those interactions are . . . historical . . . thus open to change" (Foster 2017: 131).

Although empirical and experimental scientific research on natural phenomena such as plant diseases yields useful data for conservation, historical interpretation also offers important insights (Heilbroner 1980). Some Indigenous People do not believe that their movement of pepperwood leaves for ceremonial purposes and other folk uses could contribute to disease spread, leading some people to view them as anti-science. However, the labels of primitive, superstitious, and anti-modern have been used historically to marginalize and even legally persecute Indigenous Peoples who persist in practicing their traditions, for example in the healing arts (Bauer 2016). "Scientific knowledge . . . is generated from customs and practices emanating from its particular location and history." Both Western science and Indigenous plant knowledge "are rooted in traditional pasts" and "both are dynamic as well" subject to change with new information and shifts in societal needs. Both are "simultaneously modern and nonmodern" (Foster 2017: 80–81).

According to Laura A. Foster in her book *Reinventing Hoodia: People, Plants, and Patents in South Africa*, we need to embrace systems analysis which "takes seriously the insights of feminist, postcolonial, and Indigenous theorizing by examining the intersectional politics of gender, race, indigeneity, and their colonial histories" (2017: 7). Indigenous People are both modern and traditional, still practicing their cultural traditions while also being modern political players who are "capable of negotiating their own right to self-determination" (Foster 2017: 123). For example, in the 1990s, based on well-established traditional use, various Indigenous groups and their allies successfully defended against the commercialization of ayahuasca, neem, and quinoa through patents that were later dropped (LaDuke 2005). Foster asserts that we need to "develop ways of doing science and law that place less emphasis on securing property rights and more on building meaningful collaborations that truly strengthen modes of belonging for Indigenous peoples" (2017: 130).

An international agreement signed in 1992 called the Convention on Biological Diversity attempted to establish best practices and standards for those seeking to use Indigenous knowledge to commercialize useful plants. Provisions included informed consent, sustainable use, and equitable benefit sharing (Choffnes 2016). Proponents of this approach claimed: "Drawing on ethnobotanical knowledge and compensating the providers of knowledge for their information should allow traditional societies that have conserved biodiversity to benefit from their knowledge of biological resources, thereby increasing the incentive for continuing local conservation" (Clapp and Crook 2002: 80–81). In the vanguard, Shaman Pharmaceuticals adopted these as founding principles in 1989. However, despite nearing U.S. Food and Drug Administration (FDA) approval for a promising antidiarrhea drug, they were bankrupt by 2001 in part due to the time and expense required for clinical trials. Their failure called into question whether ethnobotanical bioprospecting could compete against more efficient drug development strategies rooted in biotechnology (Clapp and Crook 2002). In addition, "not all healing plants lead to lucrative pharmaceuticals" (Osseo-Asare 2014: 208).

Based on past failures in benefit-sharing with Indigenous groups through new patent drugs, underregulated dietary supplements such as those marketed as "nutraceuticals" and "superfoods" appeared to offer a potentially more viable way to commercialize valuable plants based on traditional knowledge (Clapp and Crook 2002). However, "the weaker patent protection available to dietary supplements" carries its own risks of being undercut by the competition and vulnerable to competing claims (Clapp and Crook 2002: 119–120).[15] Despite the many challenges,

> broad policy changes over the past two decades have been in reaction to several centuries of open access to biological property for those with political power, especially under colonial regimes when there was little accountability to communities where plants were harvested.
>
> (Osseo-Asare 2014: 201)

In reality, "[m]any different institutional approaches and alliances are needed to realize the value of biological diversity and to capture the elusive incentives for ecosystem conservation that drug discovery may provide" (Clapp and Crook 2002: 95). An alternative model is to combine conservation efforts with government funded economic development projects focused on community health initiatives and female empowerment (Clapp and Crook 2002; Nightingale 2006). However, feminist economist Nancy Fraser warns against "our dangerous liaison with marketization" because marketization "liberate[s] buying and selling from moral and ethical norms" (2020: 317, 328). According to Fraser, neoliberalism depends on exploited "women's waged labor and seeks to disembed markets from democratic political regulation in order to operate all the more freely on a global scale" (2020: 303). Successful reforestation efforts led by Wangari Maathai in Africa that address various environmental issues while promoting democratic participation, including of women, offer an alternative model (Holl 2020).

Based on over a century of experimentation in Mexico, community-based management of local forest resources "can be as effective as public protected areas in conserving forest cover and biodiversity, while also generating income for local communities" (Bray 2020: 246). At the beginning of the 20th century, the federal government granted common property including forested lands to landless farmers and Indigenous Peoples, a choice rooted in "the revolutionary ideal that communities should control their own forest resources" (Bray 2020: 27). Associated rural economic development efforts work to reduce "the deleterious impacts of global capitalism" so that they can "engage with national markets . . . without being exploited" (Bray 2020: 28–29). Key to the success of Mexico's community forest enterprises is "consistent public policy support", functional and legally binding governance structures, and the collective ability of a community "to protect its interests and expand its economic and political space" (Bray 2020: 229). Community-based ownership of land with forest products which fetch high prices can increase resiliency when external regulations and community expectations give rise to sustainable forest stewardship and a "community culture of conservation" (Bray 2020: 229). Mexico's failures and successes offer lessons for community-controlled forest reserves elsewhere, including in North America.

Recent Indigenous land reacquisitions in northern California offer hope for tribes seeking to reclaim leadership roles in pepperwood conservation efforts. As a cultural keystone species, pepperwood plays roles no other species can in Indigenous societies, increasing the drive to conserve it (Garibaldi and Turner 2004). The rise of the locavore food movement in the late 20th century has renewed interest in native California foods, which are poised for greater recognition of their value given their novel flavors, adaptation to drought, and appeal to those concerned about food justice and healthy diets. There is a growing awareness in California that the ongoing process of colonialism alters "species composition and ecological structures" (Norgaard 2019: 20). For example, fire suppression caused both the density of pepperwood to increase and their productivity to decline, while global trade increased the rate of novel plant disease introductions, setting the stage for landscape-scale epidemics such as sudden oak death. Efforts to reverse these trends through ecological restoration that embraces fire as a valuable management tool holds the possibility of increased bay nut production in the future, while yet again reasserting Indigenous futures despite the "hegemonic social discourse of extermination" (Norgaard 2019: 190).

Conclusions

Beginning in the 1790s, pepperwood and its various virtues became the subject of scientific study in the universalizing Western European tradition. Limited commercial use of its leaves began by the late 19th century, continuing haltingly throughout the 20th century. Although not significantly profitable to date, commercial medicinal use of pepperwood leaves and bark was inspired by ethnobotanical research. That research bolsters contemporary Indigenous claims to property rights. In 2007, the United Nations adopted a "Declaration on the Rights of Indigenous Peoples" which states: "Indigenous peoples have the right to maintain, control, protect and develop their . . . traditional knowledge" of plants as well as "their intellectual property" (United Nations 2008: 11). Although not legally binding, the declaration is a formal recognition of the ongoing harm caused by biopiracy and the ethical imperative to do otherwise.

However, the risks persist. Based on their promising research findings, four scientists published preliminary results in 2015 on the effectiveness of pepperwood leaves and bark against antibiotic-resistant Methicillin-resistant *Staphylococcus aureus* (MRSA) to protect their legal rights to patent future medicine derived from their discoveries (Carranza et al. 2015).

Contemporary efforts to commodify bay nuts focus on marketing them as a nutrient-dense superfood. These efforts raise the possibility of recapitulating the history of valuable Indigenous food plants being culturally appropriated for profit at the expense of Indigenous people. According to sociologist Allan Johnson, the "dominant group tendency [is] to co-opt and take over anything of value produced by subordinate groups", such as wild rice and quinoa (2014: 120). Many Indigenous People seek to revitalize Indigenous foods to improve food sovereignty and health outcomes. Although selling these foods poses risks and raises complex ethical dilemmas given their sacred nature, commercial ventures rooted in traditional knowledge can result in positive economic outcomes for tribes. However, Foster succinctly summarizes the challenges of using capitalist ventures to promote social justice when she says,

> efforts to decolonize knowledge must also address how relations of power are constituted through the governing of matter and knowledge in new ways. The mere recognition or inclusion of indigenous knowledge will not be sufficient to unsettle those relations in today's market-based hierarchies that differentially value certain ways of knowing and being over others, nor is a truly just and prosperous nonracial future likely to be forged within scientific and legal practices that remain tied more to market than social justice principles.
>
> (2017: 131–132)

Despite the many challenges, contemporary efforts to commercialize bay nuts by Indigenous Peoples illustrate the concept of survivance which celebrates survival and resistance; it "accentuates the degree to which Indigenous peoples and communities have retained cultural and spiritual values and authenticity in resisting colonialism" (Smith 2012: 146). Bioprosperity offers communities the hope of crafting equitable relationships that foster sustainable use of plants, drawing from their biocultural wealth "through their own initiatives" (Osseo-Asare 2014: 202).

Acknowledgements

The following institutions graciously provided access to their collections: The Evergreen State College Library, Grace Hudson Museum and Sun House, Humboldt State University Library and Special Collections, The Huntington Library, Oregon Historical Society Museum, Oregon State University Special Collections & Archives Research Center, and the University and Jepson Herbaria Archives housed at the University of California, Berkeley. I am grateful to my students and colleagues at The Evergreen State College, with whom I have explored, in various learning communities, the history of human relations with plants. I am also thankful for a sponsored research grant during the summer of 2017, which was funded by The Evergreen State College Foundation. Anika Goldner, Tari Gunstone, and Milo Patterson assisted with the literature research. Kristina Ackley, Geena Barker, Tamara Caulkins, Susan Frankel, Jeanne Hahn, Melissa K. Nelson, and Jeffrey Wall provided helpful feedback on earlier drafts. I also thank Susan Frankel and M. Kat Anderson for independently suggesting I study pepperwood.

Notes

1 For good maps of the expedition's route, see Groves (2001).
2 For a detailed description, see Spongberg (1975).
3 According to Charles Sprague Sargent, *Umbellularia californica* "is one of the stateliest and most beautiful inhabitants of the North American forests, and no evergreen tree of temperate regions surpasses it in the beauty of its dark dense crown of lustrous foliage and in the massiveness of habit which make it one of the most striking features of the California landscape and fit it to stand in any park or garden" (1895: 22).
4 Although Costanoan is used in the literature, the term Ohlone is preferred to refer to the diverse Indigenous Peoples of the San Francisco Bay and Monterey Bay areas.
5 Sherrie Smith-Ferri clarified that you only eat the fruit flesh when the outside of the fruit is purple not green; otherwise it doesn't taste good (pers. comm. 2018). She is the Curator of Education and Exhibits and former director of the Grace Hudson Museum in Ukiah, California.
6 According to Maria Trevizo (pers. comm. 2020), stories of John Hudson still persist in Ukiah, California today among some Indigenous People who characterize him as a dupe – someone you could tell anything to about plants to earn some spending money. This of course brings into question the reliability of his unpublished notes.
7 According to Sherrie Smith-Ferri (pers. comm. 2018), Hudson used his own idiosyncratic spelling for Potter Valley Pomo words. For standardized linguistic spelling for Northern Pomo used by the tribes today, refer to the Northern Pomo language learning website which can be accessed here: http://northernpomolanguagetools.com/dictionary/word.php?word=tree#717.
8 For discussion of Indigenous ecological knowledge and practices, see Anderson (2005).
9 The Potter Valley Pomo reportedly mixed pepperwood oil with shooting star (*Dodecatheon hendersonii*) leaves to treat headaches (Welch 2013).
10 For a slightly different translation, see Meyer et al. (1999).
11 Tuberculosis was the leading cause of death in the U.S. at the turn of the century.
12 Attributed to John P. Heamy's article on *Oreodaphne Californica* first published in *American Journal of Pharmacy*, vol. 105. The correct spelling of his last name is Heany. Mr. J. P. Heany graduated from the California College of Pharmacy after completing a thesis on the tree in 1874 (University of California, Berkeley School of Pharmacy 1877).
13 The Chumash did not consume the nuts according to Jan Timbrook (2007: 221).
14 Due to its status as a rare wood and the fallacious claim that the trees are only found in California, Oregon, and incorrectly the Holy Land, a polished railroad tie of myrtlewood was used symbolically to celebrate the completion of the transcontinental railroad in 1869 as preordained by God. According to Cain, it served as the last tie joining the West and East Coast of North America with the driving of the gold spikes (2003).
15 According to Foster (2017), despite benefits sharing agreements, the San peoples of Africa did not benefit to the extent predicted when Pfizer and other pharmaceutical investors discontinued their *Hoodia* research and development due to production and marketing difficulties.

References

Abrams, L. R. 1910. A Phytogeographic and Taxonomic Study of the Southern California Trees and Shrubs, reprinted from the *Bulletin of the New York Botanical Garden* 6(21). https://doi.org/10.5962/bhl.title.13128.

Abrams, L. R., and R. S. Ferris. 1944. *An Illustrated Flora of the Pacific States: Washington, Oregon, and California*, vol. 2. Stanford University Press, Stanford, CA.

Ackerknecht, E. H. 2016. *A Short History of Medicine*, revised and expanded ed. Johns Hopkins University Press, Baltimore, MD.

Adams, J. F. A. 1887. Is Botany a Suitable Study for Young Men? *Science* 9(209): 116–117, (February 4).

Adams, Jr., J. D., and C. Garcia. 2006. Women's Health among the Chumash. *Evidence Based Complementary and Alternative Medicine* 3(1): 125–131. https://doi.org/10.1093/ecam/nek021.

Anderson, M. K. 2005. *Tending the Wild: Native American Knowledge and the Management of California's Natural Resources*. University of California Press, Berkeley, CA.

Aron, S. 2015 *The American West: A Very Short Introduction*. Oxford University Press, New York.

Baker, M. A. 1981. *The Ethnobotany of the Yurok, Tolowa and Karok Indians of Northwest California*. Unpublished Master's Thesis, Humboldt State University, Arcata, CA.

Baldy, C. R. 2018. *We Are Dancing for You: Native Feminisms and the Revitalization of Women's Coming-of-Age Ceremonies*. University of Washington Press, Seattle, WA.

Barrera, A. 2002. Local Herbs, Global Medicines: Commerce, Knowledge, and Commodities in Spanish America. In *Merchants and Marvels: Commerce, Science, and Art in Early Modern Europe*, edited by P. H. Smith and P. Findlen, pp. 163–181. Routledge, New York.

Bauer Jr., W. J. 2016. *California Through Native Eyes: Reclaiming History*. University of Washington Press, Seattle, WA.

Benemei, S., G. Appendino, and P. Geppetti. 2009. Pleasant Natural Scent with Unpleasant Effects: Cluster Headache-like Attacks Triggered by *Umbellularia californica*. *Cephalalgia* [online]. https://doi.org/10.1111/j.1468-2982.2009.01988.x.

Bleichmar, D. 2006. Painting as Exploration: Visualizing Nature in Eighteenth-Century Colonial Science. *Colonial Latin American Review* 15(1): 81–104. https://doi.org/10.1080/10609160600607499.

Boericke, W. 1906. *Pocket Manual of Homoeopathic Materia Medica: Comprising the Characteristic and Guiding Symptoms of all Remedies*, 3rd ed. Boericke & Runyon, New York.

Boericke, W. 1911. Address by Dr. Boericke. *Pacific Coast Journal of Homeopathy* 22(5): 183–188. Available at https://books.google.com/books?id=bFcCAAAAYAAJ&q=Address+by+Dr.+Boericke#v=snippet&q=Address%20by%20Dr.%20Boericke&f=false. Accessed on November 1, 2021.

Bois, P. D. 1919. Le Parc du Rougas Blanc a Marseille. *Bulletin de la Société Nationale d'acclimatation de France* 66: 163–170. Société Nationale d'Acclimatation de France, Paris, FR. Available at www.biodiversitylibrary.org/page/34249830#page/694/mode/1up. Accessed on September 10, 2018.

Bowcutt, F. 2015. *The Tanoak Tree: An Environmental History of a Pacific Coast Hardwood*. University of Washington Press, Seattle, WA.

Bowcutt, F. 2022. Plants as Luxury Foods: Affordability in an Environmentally Uncertain Future. In *A Cultural History of Plants*, Vol. 6, edited by A. Giesecke and D. Mabberley, pp. 39–61. Bloomsbury, London.

Bray, D. B. 2020. *Mexico's Community Forest Enterprises: Success on the Commons and the Seeds of a Good Anthropocene*. University of Arizona Press, Tucson, AZ.

British Pharmaceutical Conference. 1875. *Yearbook of Pharmacy, Comprising Abstracts of Papers Relating to Pharmacy, Materia Medica and Chemistry Contributed to British and Foreign Journal.* London, UK.

Brockway, L. H. 2002. *Science and Colonial Expansion: The Role of the British Royal Botanic Gardens*. Yale University Press, New Haven, CT.

Cain, A. 2003. *Oregon Myrtle, Coquille River*. The Oregon History Project, Oregon Historical Society, Portland, OR. Available at www.oregonhistoryproject.org/articles/historical-records/oregon-myrtle-coquille-river/#.YM9O8S2ZPOQ. Accessed on June 19, 2021.

Carranza, M. G., M. B. Sevigny, D. Banerjee, and L. Fox-Cubley. 2015. Antibacterial Activity of Native California Medicinal Plant Extracts Isolated from *Rhamnus californica* and *Umbellularia californica*. *Annals of Clinical Microbiology and Antimicrobials* 14: 29. https://doi.org/10.1186/s12941-015-0086-0.

Casey, L. J. 1974. Myrtlewood's Uniqueness is Worth Saving. *Our Public Lands* 24(2): 21–22. Available at https://books.google.com/books?id=24_5Tjgu15MC&pg=RA1-PA21&lpg=RA1-PA21&dq=%22Myrtlewood+Uniqueness+is+Worth+Saving%22&source=bl&ots=Z1iAvfW5SW&sig=ACfU3U0G0lJ1G0WKgfjJCe1mXSvuRgzS3Q&hl=en&sa=X&ved=2ahUKEwjq2cOe0-vzAhXSMX0KHcj2AU-AQ6AF6BAgCEAM#v=onepage&q=%22Myrtlewood%20Uniqueness%20is%20Worth%20Saving%22&f=false. Accessed on October 27, 2021.

Chesnut, V. K. 1974. *Plants Used by the Indians of Mendocino County, California*. Contributions from the U.S. National Herbarium, vol. VII. Mendocino County Historical Society, Ukiah, CA (originally published in 1902).

Choffnes, D. 2016. *Nature's Pharmacopeia: A World of Medicinal Plants*. Columbia University Press, New York.

Clapp, R. A., and C. Crook. 2002. Drowning in the Magic Well: Shaman Pharmaceuticals and the Elusive Value of Traditional Knowledge. *Journal of Environment & Development* 11(1): 79–102. https://doi.org/10.1177/107049650201100104.

Clark, L. 2015. Sonoma County Bay Nuts May be Chocolate of the Future. *The Press Democrat* (December 18).

Clarke, J. H. 1985. *A Dictionary of Practical Materia Medica. B.* Jain Publishers, New Delhi.

Cobb, R. C., D. M. Rizzo, K. J. Hayden, M. Garbelotto, J. A. N. Filipe, C. A. Gilligan, W. W. Dillon, R. K. Meentemeyer, Y. S. Valachovic, E. Goheen, T. J. Swiecki, E. M. Hansen, and S. J. Frankel. 2013. Biodiversity Conservation in the Face of Dramatic Forest Disease: An Integrated Conservation Strategy for Tanoak (*Notholithocarpus densiflorus*) Threatened by Sudden Oak Death. *Madroño* 60(2): 151–164. http://doi.org/10.3120/0024-9637-60.2.151.

Coté, C. 2016. 'Indigenizing' Food Sovereignty. Revitalizing Indigenous Food Practices and Ecological Knowledges in Canada and the United States. *Humanities* 5(57) [online]. https://doi.org/10.3390/h5030057.

Crosby, Jr., A. W. 2003. *The Columbian Exchange: Biological and Cultural Consequences of 1492*, 30th anniversary ed. Praeger, London.

Dakin, S. B. 1954. *The Perennial Adventure: A Tribute to Alice Eastwood, 1859–1953*. California Academy of Sciences, San Francisco, CA.

Dippie, B. W. 1982. *The Vanishing American: White Attitudes and U.S. Indian Policy*. University of Kansas, Lawrence, KS.

Douglas, D. 1914. *Journal Kept by David Douglas During His Travels in North America 1823–1827*. William Wesley & Son, London, UK.

Eamon, W. 1994. *Science and the Secrets of Nature: Books of Secrets in Medieval and Early Modern Culture*. Princeton University Press, Princeton, NJ.

Eastwood, A. 1939. Early Botanical Explorers on the Pacific Coast and the Trees They Found There. *California Historical Society Quarterly* 18(4): 335–346. https://doi.org/10.2307/25160840.

Eastwood, A. 1945. New Varieties of Two Well-Known California Plants. *Leaflets of Western Botany* 46(6): 166–167.

Echo-Hawk, W. R. 2010. *In the Courts of the Conqueror: The 10 Worst Indian Law Cases Ever Decided.* Fulcrum Publishing, Golden, CO.

Egerton, F. N., ed. 1983. *Landmarks of Botanical History by Edward Lee Greene*, part 1. Stanford University Press, Stanford, CA.

Flora of North America Editorial Committee, ed. 1997. *Flora of North America*, vol. 3. Oxford University Press, New York.

Foster, L. A. 2017. *Reinventing Hoodia: People, Plants, and Patents, in South Africa.* University of Washington Press, Seattle, WA.

Fraser, N. 2020. *Fortunes of Feminism: From State-Managed Capitalism to Neoliberal Crisis.* London, UK.

Frederick Stearns & Co. 1903. *The Pharmaceutical Products of Frederick Stearns & Co., Manufacturing Pharmacists.* Frederick Stearns & Co., Detroit, MI. Available at https://collections.nlm.nih.gov/catalog/nlm:nlmuid-100892352-bk. Accessed on February 10, 2021.

Garibaldi, A., and N. Turner. 2004. Cultural Keystone Species: Implications for Ecological Conservation and Restoration. *Ecology and Society* 9(3) [online]. Available at www.ecologyandsociety.org/vol9/iss3/art1. Accessed on June 20, 2021.

Garriga, A. 2001. Herbs and Remedies Used by the Indians and Californios. In *Lands of Promise and Despair: Chronicles of Early California, 1535–1846*, edited by R. M. Beebe and R. M. Senkewicz, pp. 443–445. Heyday Books, Berkeley, CA.

Go, J. 2016. *Postcolonial Thought and Social Theory.* Oxford University Press, Oxford, UK.

Goldstein, J. with D. Brown. 2013. *Inside the California Food Revolution: Thirty Years that Changed Our Culinary Consciousness.* University of California Press, Berkeley, CA.

Goodrich, J., C. Lawson, and V. P. Lawson. 1980. *Kashaya Pomo Plants.* American Indian Studies Center, Los Angeles, CA.

Gould, G. M. 1900. *An Illustrated Dictionary of Medicine Biology and Allied Sciences.* P. Blakiston's Son and Co., Philadelphia, PA.

Groves, E. W. 2001. Archibald Menzies (1754–1842), An Early Botanist on the Northwestern Seaboard of North America, 1792–1794, with further notes on his life and work. *Archives of Natural History* 28(1): 71–122.

Haller, Jr., J. S. 2009. *The History of American Homeopathy: From Rational Medicine to Holistic Health Care.* Rutgers University Press, New Brunswick, NJ.

Hedrick, U. P., ed. 1919. *Sturtevant's Notes on Edible Plants, Report of the New York Agricultural Experiment Station for the year of 1919.* J. B. Lyon Company, State Printers, Albany, NY. Available at www.biodiversitylibrary.org/page/20118943#page/597/mode/1up. Accessed on September 5, 2018.

Heilbroner, R. L. 1980. *Marxism: For and Against.* W. W. Norton & Company, New York.

Heinsen, V. E. 1987. *Mission San Antonio de Padua Herbs: Medicinal Herbs of Early Day*, 3rd ed. King City, CA (publisher not identified).

Holl, K. D. 2020. *Primer of Ecological Restoration.* Island Press, Washington, DC.

Hooker, W. J., and G. A. Walker Arnott. 1840. *The Botany of Captain Beechey's Voyage*, part 9. Henry G. Bohn, London. Originally published in February – March and reprinted in 1841. Their description of *Tetranthera californica* was first published in 1833 in an earlier part of *The Botany of Captain Beechey's Voyage.*

Howard, J. L. 1992. *Umbellularia californica*. In *Fire Effects Information System* [online]. U.S. Forest Service, Rocky Mountain Research Station, Fire Sciences Laboratory. Available at www.fs.fed.us/database/feis/plants/tree/umbcal/all.html. Accessed on October 31, 2021.

Hudson, J. W. n.d. a. Unpublished linguistic manuscript by John Hudson, no page number. Collection of the Grace Hudson Museum, Ukiah, California.

Hudson, J. W. n.d. b. Unpublished linguistic manuscript, vocabulary, p. 126, [20194–20197; 20200–20202]. Collection of the Grace Hudson Museum, Ukiah, California.

Hudson, J. W. n.d. c. Unpublished linguistic manuscript, p. 80. [20207 a, b]. Collection of the Grace Hudson Museum, Ukiah, California.

Hudson, J. W. n.d. d. Unpublished linguistic manuscript, no page number. In file folder labeled "Foods, Plants, Tobacco". Collection of the Grace Hudson Museum, Ukiah, California.

Immel, D. L. 2007. *Two Research Approaches to Ecocultural Restoration in California: Experimental Reintroduction of Showy Indian Clover and Reconstructing an Ethnobotany of the Salinan Tribe*. Doctoral Dissertation, Graduate Group in Ecology, University of California, Davis, CA.

Immel, D. L. 2018. *California Laurel: Plant Guide*. U.S. Department of Agriculture Natural Resources Conservation Service, National Plant Data Center, Davis, CA.

International Union for Conservation of Nature and Natural Resources (IUCN). 2021. *Umbellularia californica* var. *fresnensis*. The IUCN Red List of Threatened Species [web page]. Available at www.iucnredlist.org/species/35156/9915484. Accessed on October 29, 2021.

Jardine, L. 1999. *Ingenious Pursuits: Building the Scientific Revolution*. Anchor Books, New York.

Jardine, N., and E. Spary. 1996. The Natures of Cultural History. In *Cultures of Natural History* edited by N. Jardine, J. A. Secord, and E. C. Spary, pp. 3–13. Cambridge University Press, Cambridge, UK.

Jepson, W. L. 1909. *The Trees of California*. Cunningham, Curtis & Welch, San Francisco, CA.

Jepson, W. L. 1910. *The Silva of California: Memoirs of the University of California*, vol. 2. The University Press, Berkeley, CA.

Johnson, A. G. 2014. *The Gender Knot: Unraveling our Patriarchal Legacy*, 3rd ed. Temple University Press, Philadelphia, PA.

Judd, W. S., C. S. Campbell, E.A. Kellogg, P. F. Stevens, and M. J. Donoghue. 2016. *Plant Systematics: A Phylogenetic Approach*, 4th ed. Sinauer Associates, Inc., Sunderland, MA.

Keeney, E. B. 1992. *The Botanizers: Amateur Scientists in Nineteenth-Century America*. The University of North Carolina Press, Chapel Hill, NC.

Koerner, L. 1996. Carl Linnaeus in His Time and Place. In *Cultures of Natural History*, edited by N. Jardine, J. A. Secord, and E. C. Spary, pp. 145–162. Cambridge University Press, Cambridge, UK.

LaDuke, W. 2005. *Recovering the Sacred: The Power of Naming and Claiming*. Sound End Press, Cambridge, MA.

Lang, F. A. 2018. Myrtlewood. In *Oregon Encyclopedia*. Oregon Historical Society, Portland, OR [online]. Available at www.oregonencyclopedia.org/articles/myrtlewood/#.YM9IPi2ZPOQ. Accessed on June 19, 2021.

Lewis, W. H. 2000. Ethnopharmacology and the Search for New Therapeutics. In *Biodiversity and Native America*, edited by P. E. Minnis and W. J. Elisens, pp. 74–98. University of Oklahoma Press, Norman, OK.

Libster, M. 2004. *Herbal Diplomats: The Contribution of Early American Nurses (1830–1860) to Nineteenth-century Health Care Reform and the Botanical Medical Movement*. Golden Apple Publications, West Lafayette, IN.

Linné, C. 1766. *Systema naturae per regna tria naturae*, 12th ed., vol. 1, part 1. Stockholm, SE. Available at https://books.google.com/books?hl=es&id=-eIVAAAAYAAJ&dq=ina

uthor%3ACarl+von+Linné+%22diurnus%2C+varians+cultura%2C+loco%22&q=Mammalia.+Primates.+Homo#v=snippet&q=Mammalia.%20Primates.%20Homo&f=false. Accessed on October 22, 2018. (For a translation, see https://en.wikipedia.org/wiki/Scientific_racism#cite_note-20).

MacGregor, J. T., L. L. Layton, and R. G. Buttery. 1974. California Bay Oil. II. Biological Effects of Constituents. *Journal of Agricultural and Food Chemistry* 22(5): 777–780. https://doi.org/10.1021/jf60195a025.

McDonell, E. 2018. *The Quinoa Boom Goes Bust in the Andes.* The North American Congress on Latin America [web page] (March 12). Available at https://nacla.org/news/2018/03/12/quinoa-boom-goes-bust-andes. Accessed on November 12, 2019.

Meentemeyer, R., D. Rizzo, W. Mark, and E. Lotz. 2004. Mapping the Risk of Establishment and Spread of Sudden Oak Death in California. *Forest Ecology and Management* 200: 195–214. https://doi.org/10.1016/j.foreco.2004.06.021.

Merchant, C. 1980. *Death of Nature.* Harper & Row, New York (reprinted in 1990).

Meyer, F. G. E., W. E. Trueblood, and J. L. Heller, eds. and trans. 1999. *The Great Herbal of Leonhart Fuchs: De Historia Stirpium Commentarii Insignes, 1542*, vol. 1. Stanford University Press, Stanford, CA.

Moerman, D. E. 1998. *Native American Ethnobotany.* Timber Press, Portland, OR.

Moerman, D. E. 2010. *Native American Food Plants: An Ethnobotanical Dictionary.* Timber Press, Portland, OR.

Monnett, H. 1857. *The magic monitor and medical intelligencer: containing wonderful and elaborate revelations concerning the following subjects – love, courtship & marriage, how to prevent an increase of family, how to cure self-abuse and its results, the detection, prevention & cure of all private diseases, etc.* H. Monnett, New York, NY. Available at https://collections.nlm.nih.gov/bookviewer?PID=nlm:nlmuid-64220980R-bk#page/116/mode/2up/search/California+Laurel. Accessed on June 20, 2021.

Moore, J. W. 2017. The Capitalocene Part I: On the Nature and Origins of Our Ecological Crisis. *The Journal of Peasant Studies* 44(3): 594–630. https://doi.org/10.1080/03066150.2016.1235036.

Moore, J. W. 2018. The Capitalocene Part II: Accumulation by Appropriation and the Centrality of Unpaid Work/Energy. *The Journal of Peasant Studies* 45(2): 237–279. https://doi.org/10.1080/03066150.2016.1272587.

Munz, P. A. 1973. *A California Flora with Supplement.* University of California Press, Berkeley, CA.

Mushita, A., and C. B. Thompson. 2007. *Biopiracy of Biodiversity: Global Exchange as Enclosure.* Africa World Press, Inc., Asmara, Eritrea.

Nassini, R., S. Materazzi, J. Vriens, J. Prenen, S. Benemei, G. De Siena, G. la Marca, E. Andrè, D. Preti, C. Avonto, L. Sadofsky, V. Di Marzo, L. De Petrocellis, G. Dussor, F. Porreca, O. Taglialatela-Scafati, G. Appendino, B. Nilius, and P. Geppetti. 2012. The 'Headache Tree' via Umbellulone and TRPA1 Activates the Trigeminovascular System. *Brain* 135(2): 376–390. https://doi.org/10.1093/brain/awr272.

Nelson, M. K. 2004. Health and Ecology of Indigenous California Foods and the Bay Nut. *The Snail – A Slow Food Newsletter*, pp. 32–33. (The defunct newsletter was produced by Slow Food USA in early 2000s which was based in Brooklyn, NY.)

Nelson, M. K. 2021. Indigenize Conservation. *Sierra* (January/February), pp. 10–11.

Nightingale, A. 2006. The Nature of Gender: Work, Gender, and Environment. *Environment and Planning D: Society and Space* 24: 165–185. https://doi.org/10.1068/d01k.

Nisbet, J. 2009. *The Collector: David Douglas and the Natural History of the Northwest.* Sasquatch Books, Seattle, WA.

Norgaard, K. M. 2019. *Salmon and Acorns Feed My People: Colonialism, Nature, and Social Action*. Rutgers University Press, New Brunswick, NJ.

Northern Pomo Online Talking Dictionary [web page]. Available at http://northernpomolanguagetools.com/dictionary/word.php?word=tree#717. Accessed on December 27, 2020.

Nuttall, T. 1842. *The North American Sylva; Or, A Description of the Forest Trees of the United States, Canada, and Nova Scotia, Not Described in the Work of F. Andrew Michaux*, vol. 1. J. Dobson, Philadelphia, PA.

Osseo-Asare, Abena Dove. 2014. *Bitter Roots: The Search for Healing Plants in Africa*. The University of Chicago Press, Chicago, IL.

Phillips, P. W. 2016. *Ethnobotany of the Coos, Lower Umpqua, and Siuslaw Indians*. Oregon State University, Corvallis, OR.

Pratt, M. B. 1922. *Shade and Ornamental Trees of California*. California State Board of Forestry, Sacramento, CA.

Raskin, J. 2012. The Forager's Dilemma: What's Right, and Wrong, with Gathering Wild Foods? *Bay Nature* (April–June). Available at https://baynature.org/article/the-foragers-dilemma/. Accessed on August 2, 2017.

Riley, G. 1999. *Women and Nature: Saving the 'Wild' West*. University of Nebraska Press, Lincoln, NE.

Sacks, K. 2013. The Product: A Portrait of Bay Laurel. *Starchefs* (June). Available at www.starchefs.com/cook/savory/product/california-bay-laurel. Accessed on July 31, 2017.

Salmón, E. 2012. *Eating the Landscape: American Indian Stories of Food, Identity, and Resilience*. The University of Arizona Press, Tucson, AZ.

Salmón, E. 2020. *Iwígara: American Indian Ethnobotanical Traditions and Science*. Timber Press, Portland, OR.

Saltonstall, F. N. 1885. Proving of *Oreodaphne californica*. *The California Homoeopath* 3: 175–176. Available at https://babel.hathitrust.org/cgi/pt?id=mdp.39015059798861&view=1up&seq=473&q1=Oreodaphne. Accessed on February 13, 2021.

Sargent, C. S. 1895. *The Silva of North America: A Description of Trees which Grow Naturally in North American Exclusive of Mexico*, vol. 7. Houghton Mifflin and Company, New York. https://doi.org/10.5962/bhl.title.46451.

Sargent, C. S. 1922. *Manual of the Trees of North America (Exclusive of Mexico)*, 2nd ed. Houghton Mifflin Co., Boston, MA.

Schenck, S. M., and E. W. Gifford. 1952. Karok Ethnobotany. *Anthropological Records* 13(6): 377–392.

Schiebinger, L. 2004. *Plants and Empire: Colonial Bioprospecting in the Atlantic World*. Harvard University Press, London, UK.

Schiebinger, L., and C. Swan. 2005. *Colonial Botany: Science, Commerce, and Politics in the Early Modern World*. University of Pennsylvania Press, Philadelphia, PA.

Senos, R., F. K. Lake, N. Turner, and D. Martinez. 2006. Traditional Ecological Knowledge and Restoration Practice. In *Restoring the Pacific Northwest: The Art and Science of Ecological Restoration in Cascadia*, edited by D. Apostol and M. Sinclair, pp. 393–426. Island Press, Washington, DC.

Smith, K. R. 2014. *Enough for All: Foods of My Dry Creek Pomo and Bodega Miwuk People*. Heyday Books, Berkeley, CA.

Smith, L. T. 2012. *Decolonizing Methodologies: Research and Indigenous Peoples*, 2nd ed. Zed Books, London, UK.

Snow, C. H. 1903. *The Principal Species of Wood: Their Characteristic Properties*, 1st ed. John Wiley & Sons, New York.

Spongberg, S. A. 1975. Lauraceae Hardy in Temperate North America. *Journal of the Arnold Arboretum* 56(1): 1–19.

Stearns, W. T. 2004. *Botanical Latin*. Timber Press, Portland, OR.

Stein, W. L. 1990. California-Laurel. In *Silvics of North America, Vol. 2, Hardwoods*, technical coordinators R. M. Burns and B. H. Honkala, pp. 826–834. Agriculture Handbook 654. U.S. Forest Service, Washington, DC.

Stillman, J. M. 1880. Ethereal Oil of California Bay Tree. *Scientific American Supplement* 235(July 3): 3749–3750.

Strike, S. S. 1994. *Ethnobotany of the California Indians, Vol. 2: Aboriginal Uses of California's Indigenous Plants*. Koeltz Scientific Books, Champaign, IL.

Sudworth, G. B. 1908. *Forest Trees of the Pacific Slope*. Government Printing Office, Washington, DC.

Swiecki, T. J., and E. A. Bernhardt. 2008. Increasing Distance from California Bay Laurel Reduces the Risk and Severity of *Phytophthora ramorum* Canker in Coast Live Oak. In *Proceedings of the Sudden Oak Death Third Science Symposium*, technical coordinators S. J. Frankel, J. T. Kliejunas, and K. M. Palmieri. Pacific Southwest Research Station General Technical Report PSW GTR-214. U.S. Forest Service, Albany, CA.

Timbrook, J. 1990. Ethnobotany of Chumash Indians, California, based on collections by John P. Harrington. *Economic Botany* 44(2): 236–253.

Timbrook, J. 2007. *Chumash Ethnobotany: Plant Knowledge Among the Chumash People of Southern California*. Heyday Books, Berkeley, CA.

Tsing, A. L. 2015. *Mushroom at the End of the World: On the Possibility of Life in Capitalist Ruins*. Princeton University Press, Princeton, NJ.

Turner, N. 2003. 'Passing on the News': Women's Work, Traditional Knowledge and Plant Resource Management in Indigenous Societies of North-Western North America. In *Women and Plants: Gender Relations in Biodiversity and Conservation*, edited by P. L. Howard, pp. 131–149. Zed Books, London, UK.

United Nations. 2008. *Declaration on the Rights of Indigenous Peoples, Article 31* (March). Available at www.un.org/esa/socdev/unpfii/documents/DRIPS_en.pdf. Accessed on July 5, 2021.

United States Department of Agriculture, Research, Education, and Economics Information System. n.d. *Enhancing Tribal Health and Food Security in the Klamath Basin of Oregon and California by Building a Sustainable Regional System*. University and Jepson Herbaria, University of California, Berkeley. Unpublished report [online]. Available at https://reeis.usda.gov/web/crisprojectpages/0230374-enhancing tribal-health-and-food-security-in-the-klamath-basin-of-oregon-and-california-by-building-a-sustainable-regional-food system.html. Accessed on September 12, 2018.

United States Senate and United States War Department. 1856. *Reports of Explorations and Surveys, to Ascertain the Most Practicable and Economical Route for a Railroad from the Mississippi to the Pacific Ocean*, vol. 4. Beverley Tucker, Printer, Washington DC, p. 133. Available at https://babel.hathitrust.org/cgi/pt?id=uiug.30112059598372;view=1up;seq=201. Accessed on October 31, 2021.

University of California, Berkeley School of Pharmacy. 1877. *Annual Announcement of the California College of Pharmacy*. F. Eastman, San Francisco.

Van Tieghem, P. 1892. *Traité de botanique. Cours complet d'histoire naturelle* series F. Savy, Paris. http://doi.org/10.5962/bhl.title.33254.

Wall Jr., J. R. 2018. *Folk Valuation of Chestnut Diversity in Turkey: Towards Livelihood-Centered Conservation*. Doctoral Dissertation, Department of Natural Resources, Cornell University, Ithaca, NY.

Weatherford, J. 2010. *Indian Givers: How Native Americans Transformed the World*. Broadway Books, New York.

Weigand, J. 2002. California. In *Nontimber Forest Products in the United States*, edited by E. T. Jones, R. J. McLain, and J. Weigand, pp. 81–86. University Press of Kansas, Lawrence, KS.

Welch, John R. 2013. Sprouting Valley: Historical Ethnobotany of the Northern Pomo from Potter Valley, California. In *Contributions in Ethnobiology*. Society of Ethnobiology, Denton, TX.

Winson, A. 2013. *The Industrial Diet: The Degradation of Food and the Struggle for Healthy Eating*. New York University Press, New York.

Wright, W. P., ed. 1902. *Cassell's Dictionary of Practical Gardening: An Illustrated Encyclopædia of Practical Horticulture for All Classes*. Cassell and Company, London. Available at www.biodiversitylibrary.org/page/23240749#page/448/mode/1up. Accessed on September 10, 2018.

5 Nyame's altar

A cultural history of the 'God tree' in Jamaica

John Rashford

Introduction

This paper presents an historical ethnobiological approach to determining the meaning of 'God tree' in Jamaica, a name earliest recorded in Jamaica for unidentified species of trees of the genus *Ficus*, but now primarily associated with the cotton tree (*Ceiba pentandra*). Figs and the cotton tree are common in Jamaica and have similar religious significance (Baker 1965). It could be argued that, in Jamaica, the name was first applied to figs and later extended to the cotton tree, and that the cotton tree only became recognized as the God tree among colonial elites of the 19th century because it was an important feature of Myal, an African influenced religious tradition that emerged as an anti-slavery movement in the 18th century (Schuler 1979). This paper shows, however, that the explanation for the God-tree name in Jamaica is equally relevant to figs and to the cotton tree, even though it was earliest recorded for figs, and it traces the origin of this naming tradition to West Africa.

Determining the meaning of God tree in Jamaica raises issues that, when pursued, reveal a widespread pattern of thought concerning the nature of spirits and their relationship to living things, particularly trees. The need to understand this pattern of thought has long been recognized in the literature. Understanding this "system of correspondences" (Abrahams and Szwed 1983) is essential to understanding the African influence on religious and folkloric traditions of the New World, especially their role in social histories of resistance to racialized economic exploitation and its associated cultural hegemony. An ethnobiological approach to this system of correspondence adds a new perspective to the old polarized debate concerning 'African survivals' in the Americas, where the sociologist, Franklin Frazer, argued that the oppressive conditions of plantation slavery largely destroyed African traditions in the New World, especially in the United States, while the anthropologist, Melville Herskovits, attempted to document substantial continuity throughout the Americas, even in the United States (Holloway 1990; Mintz and Price 1992; Burton 1997).

DOI: 10.4324/9780429320897-6

Epiphytes and the God moniker in Jamaica

Cassidy (1961) and Le page (Cassidy and Le page 1980) are most responsible for drawing our attention to this early link between Jamaica's figs, their epiphytic beginning and the name God tree. Cassidy (1961: 380) wrote: "*God-tree*, found in the books for the Indian fig, was apparently first used in Jamaica, where a small fig may grow over another tree until it entirely swallows it up." And in the *Dictionary of Jamaican English* (1980: 201) where they cite Patrick Browne (1756) and James Macfadyen (1837), Cassidy and Le Page identify "God-tree" as both a "kind of Indian fig tree" and as the cotton tree. That spiritual significance stemming from their epiphytic beginning becomes clear when we consider other plants as well as animals with "God" as a prefix in their Jamaican common names. For example, God bush in Jamaica is the parasitic shrub *Oryctanthus occidentalis* in the Loranthaceae family, widely known by the common name mistletoe, and formerly identified as *Loranthus occidentalis* (Adams 1972: 240). The mistletoe was made famous in anthropology by Sir James Frazer's (1922) classic multi-volume work *The Golden Bough* (1890), where a global comparison of mythology, folklore and literature provided the basis for his now rejected unilineal theory of the evolution of human thought from magic, to religion, to science. In 1980, Cassidy and Le Page (1980) noted that the name God bush "indicates recognition among the [Jamaican] folk of [its] . . . 'supernatural' quality because [it grows] . . . without ever touching the ground." There is no doubt, however, that God bush considered in the context of the "system of correspondence" for things with "God" as the generic term in their compound common name points to an African origin (Cassidy and Le Page 1980: 200–201). The principle of a plant growing in or from the sky being associated with God was also responsible, Cassidy and Le Page alleged, for the name "God okro," the epiphytic night-blooming cactus, also called night-flowering cereus, known in the past as *Cereus triangularis*, but identified by Adams (1972) as *Oryctanthus occidentalis*.

The origin of the name "God-bird" (*Trochilus polytmus*), Jamaica's official national bird now widely known as the doctor bird, also called the streamer-tail or long-tail doctor bird, provides an alternative origin of the God moniker in Jamaican folk biological terminology. The name doctor bird is traditionally applied to all three species of native hummingbirds and to the small colorful Jamaican tody (*Todus viridis*), which is similar in appearance to the true hummingbirds. Today, however, it is chiefly associated with the widespread and abundant *Trochilus polytmus*, Jamaica's flamboyant endemic hummingbird which is generally regarded as one of the most magnificent of the some 320 species of hummingbirds in the world.

There are several fanciful explanations for the origin of the name doctor bird such as "their supposed resemblance to old-time doctors who dressed in black" and the idea "that the erect black crest and streamer-tail resemble the top hat and long tail coats that all doctors used to wear" (Senior 2003: 238). However the truth is most closely described in Cassidy's 1953 contribution to the

Dictionary of Jamaican English (1980: 152), recorded in the northeastern parish of Hanover, but reported for other parishes as well: "dakta bod a koni bod – haad bod fi ded/*Doctor-bird is a clever bird – a bird hard to kill*." According to Cassidy and Le Page (1980: 310), "*cunny* is cunning, and with *obeahman* (euphemistic for "bush doctor" (1980: 152, 84)) suggests the great cleverness of this bird, for it is a very hard to catch." There were many reasons for Obeah practitioners to avoid being caught. Following the 1760 uprising led by the Coromantee Tacky, in which Obeah played a prominent role, Obeah was outlawed in Jamaica: "Be it therefore enacted [that] any slave who shall pretend to any supernatural power, in order to promote the purposes of rebellion, shall, upon conviction thereof suffer death, transportation, or such other punishment" (cited in Murrell 2010: 233). Madden reports: "During the twenty years subsequent to [the 1760 uprising] . . . a great many [Africans] . . . were hanged for obeah crimes." Seventy-four years after the 1760 rebellion, the Jamaican planter Gregory Lewis (1834: 94–95) reported that his agents captured an Obeah man on his estate. Since the "good old practice of burning has fallen into disrepute," wrote Lewis, "he was sentenced to be transported, and was shipped off the island, to the great satisfaction of person of all colours – white, black, and yellow." It is in the context of the Jamaica's African religious traditions and African Jamaican resistance to slavery that the names God bird and doctor bird makes sense, and we can understand the basis of the folk belief that killing a God bird brings misfortune.

The cotton tree as the God tree in Jamaica

Although 'God tree' was first reported for figs in Jamaica, the name is today most widely associated with the cotton tree. Ceiba is the native Caribbean name recorded at the time Europeans discovered the Americas (Howe 1906). Silk cotton, silk cotton tree and cotton tree are descriptive of the cotton-like fibers called floss, which are associated with the fruit's wind-dispersed seeds. The origin and meaning of the name God tree, however, is not so straightforward.

It is reasonable to assume that the cotton tree has been of interest to African Jamaicans from early on in the island's colonial history, given its magnificence and its cultural significance in Africa. But its importance for African Jamaicans only came to the attention of colonial elites in the 19th century because of its prominence in the anti-slavery religious movement called Myal. The influential Jamaican planter and historian Edward Long, who was a staunch defender of paternalistic slavery, gave one of the earliest accounts of Myal in the 1760s. He described it as clandestine communities open to all who established contact with spirits by means of plant infusions, music and dance to make themselves invulnerable to the attacks of Europeans and capable of restoring the dead to life. The 18th century sources on Myal do not specify where contact with spirits occurred. According to Long, such encounters took place in houses. In the 19th and 20th centuries, however, there are many reports like Macfadyen's that identify the cotton tree as one of the places – often the preferred place – where contact with spirits occurred, the others being open fields, homes of the sick and

Figure 5.1 Boy and Silk Cotton Tree, Nassau, Bahama Islands (Jackson 1900).

graveyards. Writing of his experience as a West Indian proprietor in the early 19th century, Matthew Lewis (1834: 354) noted that African Jamaicans "were accustomed to meet together [for the ceremonial Myal dance] after nightfall, generally beneath the shadow of a cotton tree." There, the spirits of the sickly described as "spellbound," "nailed" or "held captive" by Obeah-controlled harmful spirits of the dead were freed by a process identified by the Myal tradition of the time as "shadow catching" (Banbury 1895; Emerick 1916).

The kinds of spirits associated with the God tree in Jamaica

Spirits of the dead

The fact that the cotton tree is called God tree in Jamaica and its possible links to African conceptions of God have been largely overlooked in favor of the tree's association with other kinds of spirits, especially spirits of the dead. Death spirits in Haiti are called zombies, and so great is the cotton tree's association with these kinds of spirits that in Haiti, it is called mapou zombie. In the 19th century Scoles (1885: 57–60) reported that in Guyana, the cotton tree was "sometimes" known as jumbie tree and "often" called "the devil's tree." Not only in Guyana, but in the Eastern Caribbean generally and in Jamaica to a limited extent, death spirits are also called jumbies, although in Jamaica, like other areas of the Caribbean – the Cayman Islands, Belize, Barbados and the Bahamas – death spirits are more widely referred to as duppies (Leach 196, Allsopp 1996: 208–209 + 317–319, Cassidy and Le Page 1980). Although 'duppy tree' as a name for the cotton tree does not appear in the early accounts of Jamaican folklore and was not recorded in the *Dictionary of Jamaican English* (Cassidy and Le Page 1980), it is noteworthy that it now occurs in newspapers and it is the title of a painting of the cotton tree by the well-known Jamaican artist, Everald Brown.

Merriam-Webster's online dictionary (2022) correctly defines duppy as "a haunting spirit of the dead" in folklore of "West Indians" but this is only one aspect of the general view in Jamaica. We get some sense of the rich connotations of "duppy" in Jamaican culture, especially its association with harm, by noting that it is also the generic term in the compound common names for a wide variety of things, including many plants and animals that are uncanny in some way, such as curiously imitative, or that are strange, eerie, dangerous or deadly (Perkins 1969; Rashford 1984, 1985, 1989). Duppies in Jamaica, then, are generally conceived of as harmful death spirits, especially the spirits of those who were not buried or who were improperly buried, as well as those who were malevolent in life. Traditionally, the latter includes dishonest shopkeepers; those who deal with death or who shed blood, such as butchers; and those who have died prematurely from accidents, suicide or homicide, especially children. It is precisely because duppies are usually thought of as harmful death spirits that they are feared, and there are well-known means of defense against them.

Duppies are said to reside at or in the canopy of the cotton tree, but the cotton tree is not only a place of spirits, it is a place for spirits. According to Leach (1961: 211): "When a body is buried its duppy can be 'planted down' – rendered powerless to leave the coffin – by various ways. . . . Two informants from different regions told me that the limb of the cotton tree placed on the coffin would keep the duppy down. This is interesting and significant," Leach continues, "since the cotton tree is the one place where the duppies most frequently gather." The "cities" of duppies, he tells us, "are in the great silk cotton trees that abound in Jamaica" (1961: 214).[1] With respect to the cotton tree, however, the following review indicates that in Jamaica, with its strong Ashanti-Akan influence, references strongly suggest the cotton tree's association not only with inner spirits, folkloric spirits and spirits of the dead, but also with the Supreme Being.

Expanding explanations for the God-tree name for figs and the cotton tree

Not surprisingly, the historic explanation for the God-tree name being associated with figs and the cotton tree has focused on their impressive size, with Browne (1756: 110) describing the fig he had in mind as a "monstrous tree" and Macfadyen (1837: 93) describing the cotton tree in terms of "the majesty of its appearance." In this context 'God tree' can be described as God-like in the sense of representing the greatness of God and the grandeur of the cosmos. While size as an explanation is based on a natural feature, the remaining four explanations are based on the religious significance of these two trees. Granted that this is true, we are still left to wonder what sacred means and why Jamaicans conceive of these particular trees as sacred.

God and Jamaica's Ashanti-Akan connection

Ethnocentrism, with its legacy of outdated cultural evolution theories, coupled with the assumption that animism, polytheism and monotheism are mutually exclusive terms, have long paved the way for the denial of Indigenous African conceptions of God, or to such conceptions being disparagingly termed "a high god" or "a High God," rather than God, the Supreme Being (e.g., Davidson 1966: 123, Smith 1995: 17). In the early works of African scholars of diverse perspectives (e.g., Idowu 1962; Mbiti 1970, 1974), the concept of a Supreme Being is widespread in the traditional religions of Africa (Pobee 1976; Ryan 1980; Smith 1995; Nkulu-N'Sengha 2009).

Consider, for example, the Ashanti of Ghana, who are discussed here given their possible role in accounting for the name God tree in Jamaica. Alfred Ellis, in his influential 1887 book, *The Tshi-Speaking Peoples of the Gold Coast off West Africa*, denied that the Ashanti had an Indigenous concept of God, citing lack of sacrifice and priesthood dedicated to it (Cited in Rattray 1916: B). Rattray, on the other hand, felt one could "hardly allow these statements to remain unchallenged, as . . . colloquial knowledge and language of the people, "(1916: 18) suggested and entirely different reality.

> In Ashanti, in remote bush villages, buried away in the impenetrable forest, and as yet even untouched by European and missionary influence, it would seem incredible that [Ahanti had] taken such deep root as to affect their folklore, traditions, customs, and the very sayings and proverbs with which their language abounds. These proverbs and traditions, moreover, which speak of and contain references to a Supreme Being, are far more commonly known among the greybeards, elders, and the fetish priestly class themselves than among the rising younger generation, grown up among new influences and often trained in the very precincts of a mission. Fetishism and monotheism would at first sight appear to the very antithesis of each other, but a careful investigation of facts will show that here in Ashanti it is not so.
>
> (Rattray 1916: 19)

Considering this, a fourth explanation for the cotton tree being called God tree is based on the continuity of Akan tradition in Jamaica. It is widely recognized that the culture of the Akan-speaking peoples of mostly modern-day Ghana, whose small states the Ashanti united into an empire between 1680 and 1891, has profoundly influenced the development of contemporary Jamaica. On West

Africa's historic Gold Coast, of which southern Ghana is a part, is the town of Koramantse, one of the earliest British slave-trading depots, and many of the Africans shipped through this depot in the late 17th and early 18th centuries were from various Akan-speaking groups. These Africans were called Coromantee in Jamaica and they have had a major impact on the island's history and culture. The Coromantee had the reputation of being fierce fighters, which accounts for a species of "stinging ants" being called Coromantee ants and the tree *Matayba apetala* being called the Coromantee tree, described by Cassidy and Le Page (1980: 132) as the "local name for the tree . . . [that] grows in the hills and is reputed for its tough wood." The most famous leaders of armed resistance against slavery in Jamaica were identified as Coromantee and Jamaica's Maroon communities have long been associated with the Coramantees.

The Akan-Ashanti influence in Jamaica is also evident in Anancy stories, the most popular genre of folktales which tell of Jamaica's trickster spider who is also associated with proverbs and sayings, and more importantly, with a whole philosophy of life – surviving by one's wits against more physically formidable opponents. "The spider in Ashanti folk-lore," writes Rattray (1916: 73), "comes easily first as the hero in most [Ashanti] . . . animal tales." Anancy represents the power of human creativity, the distinguishing feature of the species; not surprisingly, the Supreme Being as the Creator from whom humans receive their animating inner spirit is also called "*Anase Kokŏrko*, The Great Spider" (Rattray 1916: 19). Other links to the Ashanti include funeral customs that have long been reported and names for people, plants and animals; for places such as Accompong; and for traditional dishes like fufu.

Accompong, today the principal Maroon community in western Jamaica located on the margin of the rugged forested Cockpit Country in the northern part of the St. Elizabeth parish, is named after Cudjoe's brother who was second in command to Cudjoe, leader of the western Maroons. According to Sibley (1978), the origin of the name Accompong "is said to be . . . from the Ashanti word, *Nyamekopon*, which means 'the lone one, the warrior'." She made no reference to the Supreme Being. Yet Cassidy and Le Page (1980: 3) defined Accompong as "a name for the Supreme Being rendered into English as *Accompong* and understood to be [also] a personal name." Rattray (1916: 19) in his account of the different names for God among the Ashanti wrote: "The derivation of this word [*Onyankõpọń*] as *Onyàmé nkõ pọń* (*Onyàmé*, alone, great one) seems borne out by noting the word in the Akyem dialect, where it is *Onyan-koro-pọń*, (Onyame, one, great)." To be consistent, Rattray should have capitalized the synonyms for God such as "Alone," "Great," and "Great One."

Most important for the thesis this paper presents, however, is the spiritual significance of trees in Akan culture. Trees – not ornamental flowering or foliage shrubs, herbs, or vines – were the preeminent form of growing plant in the well-swept residential courtyards and public spaces of the traditional Akan settlement environment where they were valued practically for shade and as gathering places, and inspirationally for their beauty and spiritual significance. The two important categories of trees were *Nyame dua* and the "tree of reception."

Nyame dua, translated as God tree, the God tree or God's Tree (Rattray 1916: 22, 1923: 142; Pobee 1976: 2; Prussin 1980: 58; Senior 2003: 134), is especially associated with residential courtyards and shrines, and the *gyedua* or "tree of reception" (Platvoet 1985) is associated with streets and other public spaces.

Some scholars identify *Nyame dua* with the cotton tree (Senior 2003: 134) and, based on the well-known illustration of Bowdich (1819: 308), both the *Nyame dua* and the cotton tree "can be seen in close proximity in the forecourt of the entrance façade of the Asantehene's bedroom in his palace at Kumasi" (Prussin 1980: 58). For the Ashanti and other Akan, *Nyame dua* also refers to the stylized symbolic tree represented by a post set in the ground whose top forks into three or four branches that have been cut short. *Nyame dua* is the cosmic tree of the Ashanti who emphasized its specific role as the cosmic connector. Its height takes it from the earth up to the sky, the realm of God, and this characteristic makes it the appropriate place for God's offerings, which also accounts for the tree being called God's altar. Especially common in forest clearings and similar in appearance to the cotton tree, the mature *Alstonia boonei*, called *sindru* by the Ashanti, is a fast-growing tree reaching upwards of 130 feet or more with tall narrow buttresses. In the fork of the *Nyame dua* post was placed a container variously described as a basin or pot used for offerings, for God's axe called *Nyame akuma* (Rattray (1923: 142) and for collecting rainwater for ritual use. Pobee writes (1976: 2):

> [M]ost Ashanti compounds had a triple-forked branch set upright in the ground. . . . In the forked branch stood a pot in which there was water called *Nyankonsu*, i.e., the God-water (Rattray 1923: 139–142). The branch served as an altar because occasionally sacrifices of food or palm wine were made there. Again the *Nyankonsu* was sprinkled on the household as God's blessing. Thus the phenomenon of *Onyame-dua* is evidence of both the belief in the Supreme Being and of the worship of the Supreme Being who was known as *Onyame* and *Onyankopong*. The tree was symbolic of the people's dependence on God.

Nyame dua is also the name of a popular Adinkra symbol representing God's altar as a spiritual place associated with God's presence and protection. Traditionally, these symbols were most identified with the well-known Adinkra cloth worn by the heads of kin-groups and political and spiritual leaders on such occasions as funerals, naming ceremonies, marriages and public festivities, but they have also been associated with such things as traditional architecture and gold weights. In contemporary Ghana and now around the world, they have been adopted as organizational logos and incorporated into fashionable designs, aesthetic works and personal accessories. The nearness of God for the Ashanti is evidenced by the frequent mention of God in oral traditions such as naming practices, proverbs, popular expressions, traditional beliefs and folktales. It is also evident in religious festivities and artistic expression, especially music, dance and the graphic arts. The Ashanti God-tree tradition with respect to its association with actual trees and with *Nyame dua*, both in the form of the stylized symbolic tree and the symbolic design, also indicate that the Ashanti routinely engage God as the central feature of their lives.

The association of tree form with God

The profound inspirational importance of trees to human beings worldwide, especially the cosmic tree motif, which appears in diverse religious cosmologies, has long been noted in ethnographic and archaeological studies and in the works of scholars in history, religious studies, geography, folklore, ethnobotany and the arts (Anonymous 1890; Philpot 1897; Beech 1913; Maguire 1931; Collis 1954; Condit 1969; Cook 1974; Reno 1977; Simoons 1998; Altman 2000; Jones and Cloke 2002; Anderson 2003; Kunwar and Bussmann 2006; Kislev et al. 2006). One of the most influential views of the nature of the cosmic tree was presented by the religious historian Mircea Eliade, who emphasized its essential nature as the center of the universe that was also the connection between the sky and earth. The idea of the "center" is a fundamental concept in Eliade's thinking. Summarizing his view of the cosmic tree and its relationship to the religious importance of trees generally, he wrote (1991: 44):

> The most widely distributed variant of the symbolism of the Center is the Cosmic Tree, situated in the middle of the Universe, and upholding the three worlds as upon one axis. . . . It may be said, in general, that the majority of the sacred and ritual trees that we meet within the history of religions are only replicas, imperfect copies of this exemplary archetype, the Cosmic Tree. Thus, all these sacred trees are thought of as situated in the Center of the World.

Eliade's conception has extensive ethnographic support and is widely represented in scholarly publications, especially in encyclopedias and other reference works, but it is limited. If the cosmic tree is truly cosmic, it must be both the center and the periphery, that is, it must be the connector as well as the sky and the earth that it connects.

An alternative, yet complementary, perspective not as widely discussed as Eliade's but cited by researchers for more than a hundred years is the cosmic tree as an ancient conception that models the structure and operation of the universe as an all-encompassing treelike oppositional unity. With its branches above and roots below, it is likely that the tree has been widely chosen to model the universe because it fits well the human experience of the sky above, the earth below and the surface in the middle where the sky and earth meet. The tree-modeled cosmic whole is the grounding motif of the cosmic tree, and ideas of the cosmic tree as cosmic center or cosmic connector, or as cosmic center-connector, are only its varying expressions. For a more coherent understanding of the Akan tradition of *Nyame dua*, as well as other African-derived religious traditions like Candomble in Brazil; Vodou of Haiti; Santeria of Cuba; and Obeah, Myal and Kumina in

Jamaica, the conception of a tree-like universe (that is botanically, architecturally and stylistically expressed) provides a more adequate framework than Eliade's emphasis on the cosmic tree as preeminently the cosmic center-connector (Deren 1970: 36, Desmangles 1977: 15; Rey 2005: 1659; Prussin 1999: 430).

Consider, for example, the cosmic tree of Haitian Vodou which has its botanical form associated with figs and the cotton tree, its architectural form that is the center post of their ceremonial space and its stylized forms exemplified by the verve drawings and the African and Christian crosses. The cross in Vodou, according to the main thesis of Desmangles's article *African interpretations of the Christian cross in Vodun* (1977), is of African origin and the Christian cross has essentially been Africanized:

> The cross is the most important symbol in Vodun. It represents the crossroads of human destiny as represented by Legba; it is the symbol of the avenue of communication between the profane world of men and the sacred world of the *loas*. Moreover, like the Dahomeans, Vodunists see in the cross the entire cosmos. . . . It also represents the four cardinal points of the universe, points at which the *loas* are said to reside. In short, the cross is the central symbol in Vodun. It is the basis for Vodunists' Weltanschauung. They see it as the supporting framework around which the universe is constructed.
>
> (1977: 15)

From the perspective of this paper, however, it is not the cross per se but the cosmic tree that is at the center of Haitian Vodou: the cross is only its stylized representation. According to Simpson (1945: 50), the fig is the cosmic tree of Haiti and,

> is the "abode" of Legba, the residing loa who guards the cosmic crossroads through which the loas pass to dance with their devotees. The religious significance of the fig tree in Haiti is not as well represented in the literature as the cotton tree which Rey (2005: 1659) identified as the "most venerated tree in Haitian Vodou.

The loas that are especially associated with the cotton tree include Legba as well as Gede, guardian of the crossroad of life and death, Gran Bwa, the loa of the forest and Loco, the loa whose sphere of influence includes plants indispensable for healing.

Conclusion

An ethnobiological approach to explaining the meaning of the name God tree in Jamaica supports the view that there is a "system of correspondence" associated with the relationship between nature, spirits and the realization of human desires that is essential to understanding African religious and folkloric traditions of the New World, especially their role in social histories of resistance to economic exploitation and cultural repression. The hope is that this paper will encourage a more critical approach to this system of correspondence in African influenced religious and folkloric traditions in the Americas, for as Beckwith (1929: 117) has rightly observed: "Plant life is alive with spirit power." So too is the rest of the natural world. An ethnobiological understanding the cultural significance of figs and of the cotton tree in the African Atlantic, including the meaning of the God-tree name for the Akan and for Jamaicans, is an expression of this fact.

Note

1 Beyond death spirits, the worst kinds of duppies are those of folklore, some of which are associated with the cotton tree. They include those with human form, such as the red-eye whooping boy seen and heard on moonlight nights, and the poison-breathing Bubby Susan, whose grotesquely long breasts are thrown over her shoulder when she chases her victims. These folkloric spirits also assume the form of animals, such as the murderous fire-breathing chain-dragging rolling calf, the nightmare of travelers, and the fast-galloping three-foot horse, whose hot breath kills, and who, like rolling calves, is seen on moonlight nights (Jekyll 1907; Leach 1961; Senior 2003).

References

Adams, C. D. 1972. *Flowering Plants of Jamaica*. University of the West Indies, Kingston.

Allsopp, R. 1996. *Dictionary of Caribbean English Usage*. University of the West Indies Press, Jamaica, Barbados, Trinidad & Tobago.

Altman, N. 2000. *Sacred Trees: Spirituality, Wisdom and Well-Being*. Sterling Publishing Co., Inc., New York.

Anderson, Kit. 2003. *Nature, Culture and Big Old Trees: Live Oaks and Ceibas in the Landscapes of Louisiana and Guatemala*. University of Texas Press, Austin.

Anonymous. 1890. *Cultus Arborum*. Privately printed. Available online through Kessinger Publishing.

Baker, Herbert G. 1965. The Evolution of the Cultivated Kapok Tree: A Probable West African Product. In *Economic Development in Tropical Africa*, edited by David Brokensha. University of California, Berkeley.

Banbury, Thomas R. 1895. *Jamaica Superstitions, or the Obeah Book*. de Souza, Kingston.

Beckwith, Martha W. 1969 (1929). *Black Roadways: A Study of Jamaican Folk Life*. Reprinted in Negro Universities Press, New York.

Beech, W. H. 1913. The Sacred Fig-Tree of the A-Kikuyu of East Africa. *Man* 13(3): 4–6.

Bowdich, Edward. 1819. *Mission from Cape Coast Castle to Ashantee, With a Statistical Account of that Kingdom*. John Murray Publisher, London.

Browne, Patrick. 1756. *The Civil and Natural History of Jamaica*. T. Osborne and J. Shipton Publishers, London.

Burton, Richard D. E. 1997. *Afro-Creole: Power, Opposition, and Play in the Caribbean*. Cornell University Press, Ithaca, NY.

Cassidy, Frederic G. 1961. *Jamaica Talk: Three Hundred Years of the English Language in Jamaica*. Macmillan, London.

Cassidy, Frederic G., and R. B. Le Page. 1980. *Dictionary of Jamaican English*. Cambridge University Press, Cambridge.

Collis, J. S. 1954. *The Triumph of the Tree*. The Viking Press, New York.

Condit, I. J. 1969. *Ficus: The Exotic Species*. Division of Agricultural Sciences, University of California Press, Berkeley, CA.

Cook, R. 1974. *The Tree of Life: Images for the Cosmos*. Thames and Hudson, New York.

Davidson, Basil. 1966. *African Kingdoms*. Time-Life Books, New York.

Deren, M. 1970. *Divine Horsemen: Voodoo Gods of Haiti*. A Delta Book, New York.

Desmangles, L. G. 1977. African Interpretations of the Christian cross in Vodun. *Sociological Analysis* 38(1): 13–24.

Eliade, Mircea. 1991. *Images and Symbols: Studies in Religious Symbolism*. Princeton University Press, Princeton.

Emerick, A. J. 1916. *Jamaica Mialism*. Printed privately, Woodstock.

Frazer, James George. 1922. *The Golden Bough: A Study in Magic and Religion*. Macmillan, New York.

Herskovits, Melville. 1930. Review of Black Roadways: A Study of Jamaican Folk-life, by Martha Beckwith. *The Journal of American Folklore* 43(69): 332–338.

Holloway, Joseph, E. 1990. *Africanisms in American Culture*. Indiana University Press, Bloomington.

Howe, Marshall A. 1906. Some Photographs of the Silk-cotton Tree (Ceiba pentandra), with Remarks on the Early Records of Its Occurrence in America. *Torreya* 6(11): 217–231.

Idowu, Bolaji E. 1962. *Olodumare: God in Yoruba Belief*. Longman, Ikeja, Nigeria.

Jackson, William Henry. 1900. *Ceiba or Silk Cotton Tree, Nassau, Bahama Islands*. Library of Congress, Washington, DC.

Jekyll, Walter. 1907 (1966 edition). *Jamaica Song and Story: Annancy Stories, Digging Sings, Ring Tunes, and Dancing Tunes*. Dover Publications, Inc, New York.

Jones, Owain, and Paul Cloke. 2002. *Tree Cultures: The Place of Trees and Trees in their Place*. Oxford, New York.

Kislev, M. E., A. Hartmann, and O. Bar-Yosef. 2006. Early Domesticated Fig in the Hordan Valley. *Science* 312: 1372–1734.

Kunwar, R. M., and R. W. Bussmann. 2006. Ficus (Fig) Species in Nepal: A Review of Diversity and Indigenous Uses. *Lyonia* 11(1): 85–97.

Leach, MacEdward. 1961. Jamaican Duppy Lore. *Journal of American Folklore* 74: 207–215.

Lewis, Matthew Gregory. 1834. *Journal of a West Indian Proprietor, kept during a Residence in the Island of Jamaica*. John Murray, Albemarle Street, London.

Macfadyen, James. 1837. *The Flora of Jamaica Volume 1; A Description of the Plants of that Island, Arranged According to the Natural Orders*. University of California Press, Berkeley, CA.

Maguire, R. A. J. 1931. Masai Water-Trees. *Man* 31(July): 142–143.

Mbiti, John S. 1970. *Concepts of God in Africa*. SPCK, London.

Mbiti, John S. 1974. Traditional Religions in Africa. In *Historical Atlas of the Religions of the World*, edited by Isma'īl Rāgī al Fārūqī. Macmillan Publishing Co., Inc., New York.

Merriam-Webster Online Dictionary. 2022. *Duppy*. Available at https://www.merriam-webster.com/dictionary/duppy.

Mintz, Sidney, and Richard Price. 1992. *The Birth of African-American Culture: An Anthropological Perspective*. Beacon Press, Boston.

Murrell, Nathaniel Samuel. 2010. *Afro-Caribbean Religions*. Temple University Press, Philadelphia.

Nkulu-N'Sengha, Mutombo. 2009. God. In *Encyclopedia of African Religion*, edited by M. K. Asante and A. Mazama. A Sage Reference Publication, Los Angeles.

Perkins, Lilly. 1969. Duppy Plants in Jamaica. *Jamaica Journal* 3(10): 17–21.

Philpot, J. H. 1897. *The Sacred Tree in Religion and Myth*. Macmillan and Co., Limited, London.

Platvoet, J. G. 1985. Cool Shade, Peace and Power: The Gyedua ('Tree of Reception') as an Ideological Instrument of Identity Management among the Akan Peoples of Southern Ghana. *Journal of Religion in Africa* XV(3): 174–200.

Pobee, John. 1976. Aspects of African Traditional Religion. *Sociological Analysis* 37(1): 1–18.

Prussin, Labelle. 1980. Traditional Asante Architecture. *African Arts* 13(2): 57–65+78–82+85–87.

Prussin, Labelle. 1999. Non-Western Sacred Sites: African Models. *The Journal of the Society of Architectural Historians* 58(3): 424–433.

Rashford, John. 1984. Plants, Spirits and the Meaning of 'John' in Jamaica. *Jamaica Journal* 17(2): 62–70.

Rashford, John. 1985. Cotton Tree and the Spiritual Realm in Jamaica. *Jamaica Journal* 18(1): 65–72.

Rashford, John. 1989. Barringtonia. *Jamaica Journal* 22(2): 65–72.

Rattray, Sutherland R. 1916 (1923). *Ashanti Proverbs*. Clarendon Press, Oxford.

Rattray, Sutherland R. 1923 (1969). *Ashanti*. Clarendon Press, Oxford.

Reno, S. J. 1977. Religious Symbolism: A Plea for a Comparative Approach. *Folklore* 88(1): 76–85.

Rey, T. 2005. Trees in Haitian Vodou. In *The Encyclopedia of Religion*, edited by B. R. Taylor and J. Kaplan, vol. 2, pp. 1658–1659. Thoemmes Continuum, New York.

Ryan, Patrick J. 1980. 'Arise, O God!' The Problem of 'Gods' in West Africa. *Journal of Religion in Africa* XI(3): 161–171.

Schuler, M. 1979. Myalism and the African religious tradition in Jamaica. In Margaret E. Crahan and Franklin W. Knight (Eds.), *Africa and the Caribbean: The legacies of a link*. Baltimore: The Johns Hopkins, 65–88.

Scoles, Reverend J. S. 1885. *Sketches of African and Indian life in British Guiana*. Demerara, Georgetown.

Senior, Olive. 2003. *Encyclopedia of Jamaican Heritage*. Twin Guinep Publishers, St. Andrew, Jamaica.

Sibley, Inez Knibb. 1978. *Dictionary of Place-names in Jamaica*. The Institute of Jamaica, Kingston.

Simoons, F. J. 1998. *Plants of Life, Plants of Death*. The University of Wisconsin Press, Madison.

Simpson, George E. 1945. The Belief System of Haitian Vodun. *American Anthropologist* 47(1): 35–59.

Smith, Jonathan Z., General ed. 1995. *The HarperCollins Dictionary of Religion*. Harper Collins Publishers, San Francisco.

6 "Paddle with the tide"

Nuu-Chah-Nulth and western redcedar coevolution

Marie J. Zahn, Joe Martin and Eli Enns

Introduction

> The cedar tree is known by Nuu-chah-nulth peoples as *ḥumiis*. This is simply because every aspect of Nuu-chah-nulth life is supported by the cedar elder plant relatives. The cedar trees are thought of as elder relatives for several reasons, including its long lifespan, physical stature, presence in deep historical story and information, as well as its healing and transformative properties. An incredibly dynamic life force, some modern folks have referred to cedar as the tree of life, which is very appropriate all things considered. Taking a *quay'qwiiksup* perspective on *ḥumiis* suggests that it is in an advanced state of evolution for the climate of Nuu-chah-nulth and may be vulnerable in the face of climate change due to mass over harvest and habitat disruption. *Quay'qwiiksup* also suggests that co-evolutionary co-dependence with salmon and other species will further undermine *ḥumiis's* ability to survive in its current state. Following *iisaak*, Nuu-chah-nulth culture would have to adapt and evolve along with these species and any of their possible successors.
>
> (Eli Enns 2021)

As the world reaches a bioclimatic threshold where human activities are harming nature at unprecedented rates, a new moral framework is gravely needed where the natural world is conserved for its own sake. Knowledge of the harmful impacts caused by anthropogenic climate change and biodiversity loss has led to alarming narratives and polarizing perspectives about the state of the planet. Väliverronen (1998) beautifully explains the integral role that metaphors have for people to understand and communicate complex issues like climate change. Metaphors not only serve as a comprehension tool but can also form a powerful bridge between biology and ethics (Weingart 1995; Väliverronen 1998). For example, conservation biology has become known to be a "crisis discipline" where its research is often shaped by ethical concerns surrounding ecological preservation and sustainability (Meffe 1998; Väliverronen 1998). As such, conservation biology has departed from being a purely scientific discipline and has largely developed into an emblem of the environmental crisis narrative. Growing scientific evidence for environmental change has led to the use of strong metaphors such as the "war on climate change" or "sixth mass extinction" (Ceballos et al. 2015) to express its large scale and the emotions felt for

DOI: 10.4324/9780429320897-7

this topic. Importantly, these metaphors and perspectives also shed light on the underlying values people have about the natural world.

As long-time residents of traditional territories, Indigenous communities have witnessed and experienced the effects of environmental change and species decline first-hand. All over the world, Indigenous People have noted these changes and expressed their concern for the conservation of their lands for future generations (e.g., Turner and Clifton 2009). However, as shown for Indigenous communities in the Arctic, climate change is not necessarily their primary concern as they fight for cultural preservation, political autonomy, equity, and decolonization (Huntington et al. 2019). For Arctic Indigenous communities and elsewhere, environmental change is not described using adverse terms like "war" or "climate apocalypse" in a doom and gloom perspective, instead, it is contextualized by emphasizing their personal relationship with the environment marked by long-term resilience and adaptation (Huntington et al. 2019). When climate change is framed only using alarming images such as an emaciated polar bear searching for food, it reinforces human colonial power over natural systems. These metaphors emphasize people's control and management over the environment, as opposed to fostering a human-nature relationship bounded by respect. Further, they ignore local perspectives of resident Indigenous Peoples who have lived in close relationship with their environment and have proven to be adaptable to climate changes (Liverman 2009; Cameron 2012; Huntington et al. 2019).

In the following pages, we explore Pacific northwest coast Indigenous perspectives that serve to shift people's values away from the dominion of nature and toward a profound respect for it. We explore the Nuu-Chah-Nulth First Nation worldview of *heshook-ish tsawalk*, meaning "everything is one and everything is interconnected" by discussing relationships and practices with western redcedar (*Thuja plicata*), a cultural keystone species for Pacific northwest coast Indigenous communities (Garibaldi and Turner 2004; Zahn et al. 2018). We delve into the Nuu-Chah-Nulth cultural obligation to observe, appreciate, and act accordingly and how this worldview exemplifies a social-ecological system where people and cedar continue to coevolve despite the prevailing changes of biodiversity loss and a changing climate. Elders teach their children to find strength in change and to not fight it: "Paddle with the tide, not against it" (Eli Enns 2019). In contrast to the crisis narrative discussed previously, this chapter offers an alternative perspective to climate change that is characterized by resilience. Just like their redcedar neighbors, Nuu-Chah-Nulth individuals embody what it means to be resilient to modern-day changes and understand that this relationship is requisite for effective adaptation to an ever changing and impermanent environment.

Methods and positionality statement

This chapter focuses on the knowledge and experiences of Eli Enns and Joe Martin, both members of the Tla-o-qui-aht First Nation that is part of the broader Nuu-Chah-Nulth community along the west coast of Vancouver Island, British

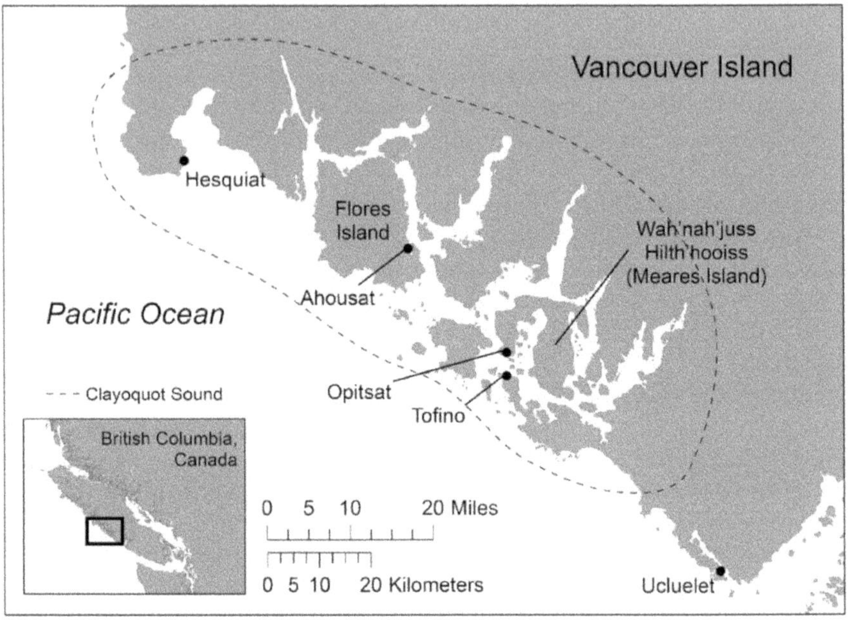

Figure 6.1 Clayoquot Sound, the sovereign territories of the Nuu-Chah-Nulth nation located on the northwest coast of Vancouver Island, British Columbia in Canada.

Columbia, Canada (Figure 6.1). Eli Enns is from Tla-o-qui-aht on his father's side and Dutch Mennonite on his mother's side. As a political scientist, Eli has studied international dispute resolution and constitutional law and is a leader for a number of initiatives: CEO and president of the IISAAK OLAM foundation, regional coordinator of North America for the Indigenous Peoples' and Community Conserved Areas and Territories (ICCA) Consortium, and CEO and president of the CleanTech Community Gateway. Joe Martin, also known by his traditional name Tutakwisnapšiƛ, is an elder in his community and is known for his expertise as a world-renowned master canoe carver. His father taught him how to hunt, trap, and build canoes, much like Joe's father had learned these skills from his father. Both Eli Enns and Joe Martin are active members and teachers in their community, revitalizing ancestral teachings and advocating for their native lands.

Marie J. Zahn is a non-Indigenous scientist and ally who led the cocreation of this chapter with Eli and Joe. Together, the authors fostered an ethical space where two knowledge systems – Western science and Indigenous knowledge – came together as equals, as opposed to occasions where one dominates the other. Here, Eli and Joe contribute their depth of knowledge to show how their experiences and relationships with western redcedar demonstrate Nuu-Chah-Nulth values. Marie conducted semi-structured interviews with Eli and Joe and subsequently analyzed their transcripts for apparent themes and notable accounts. As recorder, writer, and ally, Marie served to transfer Eli and Joe's perspectives into written form.

Cedar in Nuu-Chah-Nulth culture

First Nation relationships with western redcedar in coastal British Columbia (BC) exemplify human-nature interconnectivity. Botanically, western redcedar is a large evergreen, coniferous tree that currently ranges from southeast Alaska to northern California on the Pacific coast. The oldest individuals live well over 1,000 years, and even after death, this cedar can persist for years due to a rot-resistant oil called *thujaplicin* in its wood that prevents decay (Wilms and Turner 2014). Well after its biological death, western redcedar continues to provide habitat and nutrition for neighboring species as a snag or nurse log. Western redcedar has been identified as a cultural keystone species for Pacific Northwest Indigenous Peoples as it has been central to their cultural identity and material needs for thousands of years (Garibaldi and Turner 2004). Today, western redcedar is colloquially referred to as 'cedar,' but depending on the context, cedar may refer to either redcedar or yellow cedar (*Cupressus nootkatensis*). To Nuu-Chah-Nulth peoples, the cedar tree is known as *ḥumiis* (pronounced 'who-miss') and hereafter, we refer to the tree by this name.

The uses for *ḥumiis* by Indigenous communities are vast and deeply historied, having spiritual, economic, and ecological importance. From boxes and woven baskets to dugout canoes and building materials, every part of western redcedar has been used (Stewart 1995; Garibaldi and Turner 2004). *Ḥumiis* products, like canoes and large wooden planks, were also a commodity to be traded for various supplies with other tribes along the coast. *Ḥumiis* maintains a strong presence beyond material applications as it is at the core of numerous First Nation spiritual ceremonies and medicines (Stewart 1995; Zahn et al. 2018). Undeniably, western redcedar is woven into the physical and spiritual fabric of Indigenous culture and livelihood. Outside of Indigenous customs in BC, western redcedar has been significant for the forestry industry because its rot-resistant wood is commercially valuable. The wood is suited for a wide range of applications, most commonly for fences, decks, siding, and power poles (Pojar and MacKinnon 1994; Wilms and Turner 2014). Typically, commercial forestry harvests *ḥumiis* through clear-cutting where most, if not all, trees are logged in a given area. Clear-cut logging is one of several major threats to Pacific Northwest forests, the others being development and fragmentation, pest outbreaks, and climate change (Jakob 2000; Metsaranta et al. 2011; Murdock et al. 2013).

Central to Nuu-Chah-Nulth teachings of *hešhook-ish tsawalk* is respect and that to be respectful is to observe, appreciate, and act accordingly (Atleo 2004, 2011; Eli Enns, pers. comm. 2019). Nuu-Chah-Nulth people respect nature no differently than respecting another person. Yet, the teaching goes beyond the understanding of interconnectedness between humans and their environment; it instills the principles of how to interact with one's environment. First, people need to listen and bear witness to themselves and their surroundings.

This observation requires putting the individualistic ego aside and accessing a level of awareness both internally and externally that is present and vulnerable. A natural outcome from these observations is sincere appreciation and valuing the existence and life of nature for its own sake, not for its human services. Finally, action is taken in accordance with these mental states. Typically, this is seen by only taking what is needed (e.g., a strip of *ḥumiis* bark or piece of a root) and providing reciprocity of some kind (e.g., fish bones or shells).

Heshook-ish tsawalk is exemplified in numerous ways with *ḥumiis*, but one of the most commonly described is during its harvest. As a master carver, Joe Martin intimately knows and teaches how he selects and fells an old growth *ḥumiis*. He first visits a place several times to listen, pray, and share food, observing his needs from the forest. Joe and Eli explained how they ask permission from Great Spirit to have the tree. Eli explained,

> You don't cut down a tree and then figure out what products you're going to make after that. You come with a clear intention; you need to have a need in mind, then you go to the forest to ask for that need.
>
> (Eli Enns 2019)

In the process to select a tree, Joe further described:

> My dad and my grandfathers taught me that it is never good to cut a tree down in the spring or the summertime. Because those birds [sparrows] could be nesting up there and we have respect for those creatures. The only time trees were felled was in the fall or wintertime when those birds are no longer nesting. Today, people cut trees any old time and don't pay attention – a huge disrespect for nature. We're taught to be aware of everything.
>
> (Joe Martin 2019)

He takes careful note of his surroundings, selecting a tree that does not house a bird nest and does not have a cougar, bear, or wolf den nearby. The practice of harvesting *ḥumiis* includes acute awareness of its impacts, ensuring that it does not harm other species also living in the forest and to only take a minimum, leaving something in return. Culturally modified *ḥumiis* are a notable example where Indigenous People only took what they needed (e.g., a single plank or strip of bark) without felling the entire tree (Mobley et al. 1992; Turner et al. 2009).

Nuu-Chah-Nulth values extend through every step as they engage with *ḥumiis* from the harvest of its boughs, wood, and other materials, to the uses of the tree that extend to both everyday and spiritual realms. Of particular importance is the *ḥumiis chee-nulth*, or cedar totem pole in English, that captures ancient symbols and cultural narratives through imagery masterfully carved into the wood. As white colonists arrived to Nuu-Chah-Nulth territory, *chee-nulth* were in front of every house. The first Europeans to arrive to Opitsaht (Figure 6.1) stated Indigenous Peoples were illiterate to their English writing, but so too were the colonists to the Indigenous Peoples' *chee-nulth*. As a master carver himself, Joe Martin explained the primary components carved into a cedar *chee-nulth*:

> Normally people will see on top of a totem pole a thunderbird with wings open. Those are about our male ancestry. The ones with the wings closed is about our female ancestry. And the other thing you'll find on the top of a totem pole would be the crest of the sun or the moon. The arrangements of the crest on there were very important. It's placed on top because it's about the first teachings, the first law, and that is about respect . . . that's a teaching that carried on since you were conceived all the way until you die. If it's not on top of the totem pole, it'll always be on the chest of the thunderbird, the male or female, always on the chest because it's about the first teaching and first law of respect. . . . On the bottom of a totem pole is either a wolf, a bear, or the killer whale. And they are not on the bottom because they are the bottom – they are the ones that uphold natural law.
>
> (Joe Martin 2019)

Joe further explained the ethos of some of the totems carved into the *ḥumiis* and related initiation events. As one of the totems situated at the bottom, the wolf upholds important foundational teachings within Nuu-Chah-Nulth culture. Joe continued:

> When you lose all your baby teeth it was time for the initiation into the wolf clan, and that takes place in the wintertime, after December 21st. . . They'd wait for the new moon and they'd commence. The reason for this initiation into the wolf clan was – it was a very significant thing – it was about the second most important teaching of life, which is about fear. Never have fear or show fear because the elders said if you wanted to learn something (they'd put their fingers out, half an inch apart), you'd only learn this much. If you're not afraid, they'd spread their arms wide open and say you can learn anything. That was the beginning of the teachings for the rest of your life about all your roles and responsibilities pertaining to the village and also to nature. The Elders always said Mother nature will provide for our needs, but not our greeds.
>
> (Joe Martin 2019)

Fundamental Nuu-Chah Nulth teachings, including those about respect and fear, were carved into *ḥumiis chee-nulth* that were raised as monumental pillars. The position of a *chee-nulth* in front of each person's home demonstrated that every one of those people understood and embodied the teachings of natural law. *Chee-nulth* continue to serve this purpose, but are now placed in more public, urban centers where non-Indigenous audiences can learn about Nuu-Chah-Nulth traditions.

For thousands of years, Nuu-Chah-Nulth people and Indigenous communities of the Pacific Northwest have continued to transfer their ancestral knowledge through art, song, dance, and oral narratives. Joe Martin described growing up learning from his father, trapping, hunting, and fishing; he said, "There was no classroom. Nature was our classroom." In addition to totem poles, cedar assumed a central role in Nuu-Chah-Nulth cultural settings and was incorporated into important initiation events for young individuals reaching adulthood. Eli Enns shared one such initiation event:

> When a female child becomes a woman, she would be put away from the community on the first moon cycle. She would be put away from the community and no one would see her except the elder women in her family. She would receive a very tailored education at that time from the older women in her family about womanhood and about what her roles and responsibilities are as a woman in the community. Then after a given period of time, it would be three months or four months or six months, depending, she would be brought back out into the community ceremoniously, and she would be carried in a cedar canoe. She would be carried into a cedar longhouse and she would be carried by the men of her family but not one of them would touch her. No male then would be permitted to lay hands on her until she has a husband.
>
> She would be brought into the center of the longhouse where the floor has been opened. Any business of the nation that needs to be done would happen in the open floor, in the open space. She would be brought in and they would turn her as they move into the space and then while she was in that space she would be ceremoniously given a new name. She would no longer carry the name of her childhood. She would be given a new name of her womanhood. Then the men in her family would ceremoniously wash her feet in a basin of water in a cedar basin. They would pour the water over her feet without touching her feet and they would collect that water. . . They would pour that water into the stream and they would introduce this young woman back to the world. This is about maintaining relationship and about *heshook-ish tsawalk*.
>
> (Eli Enns 2015)

Clearly evidenced by Eli's description, *ḥumiis* made numerous appearances in the ceremonial initiation of a young girl to womanhood. Cedar was an important building material and article in sacred contexts, woven into the daily and spiritual lives of Nuu-Chah-Nulth people. Stemming from this interwoven relationship with *ḥumiis* and other life forms, Nuu-Chah-Nulth individuals embodied their teachings of respect and reciprocity. *Heshook-ish tsawalk* ensures harmony is maintained between the human and non-human worlds because they are not seen as separate.

Cedar coevolution

Like its salmon and fern neighbors, *ḥumiis* has evolved in harmony with its ecological surroundings where complex, interconnected ecological processes have shaped its existence. A notable example is when anadromous salmon return to their natal streams to spawn, and upon their death, their carcasses enrich the surrounding riparian habitat with essential nutrients (Reimchen et al. 2002; Drake et al. 2006). Nitrogen isotope tracers that were introduced to a forest with a salmon-bearing stream showed a rapid uptake of ^{15}N into the roots of western redcedar within a week (Reimchen et al. 2002). However, *ḥumiis* has not only evolved from a rich ecological landscape, but it has also emerged from the cultural landscapes and livelihoods of the resident peoples. For Indigenous communities like the Nuu-Chah-Nulth, the coevolution between *ḥumiis* and people has been defined by respect and reciprocity. Over the last century, anthropogenic emissions, commercial development, and other activities have caused pronounced ecological changes that have disrupted the harmony that was present in pre-industrial times. In BC, the temperature is expected to increase by 1.3–2.7°C by 2050 (BC Ministry of Environment 2016). From 1900–2013, the province has warmed at a rate of 1.4°C per century, surpassing the global average of 0.85 per century (BC Ministry of Environment 2016). Moreover, global mean land surface air temperature has risen to nearly twice that of the global average temperature (land and ocean) from pre-industrial times (1850–1900; IPCC 2019). Anticipated drier summers present risks for numerous species, but western redcedar in particular is expected to experience stress and likely contract its distribution and shift northward to more suitable cooler climates (McKenney et al. 2011; Case and Lawler 2017). BC residents, biologists, and Indigenous individuals have observed more severe and frequent flagging of *ḥumiis* trees (Zahn et al. 2018). Flagging occurs when a soil-moisture deficit causes leaves to turn shades of yellow and brown before shedding; if severe enough, the tree may die. Western redcedar needs a moist environment and it has been considered a "canary in the coalmine," or indicator species for a forest; when redcedar shows signs of stress, it often indicates poor ecosystem health (Zahn et al. 2018).

Climate change is one of the most pressing biopolitical debates at present and has fueled strong – and sometimes antagonistic – responses to find solutions quickly. Eli Enns explained that the present-day "war on climate change" is only a continuation of Western philosophy that led to what has become known as climate change in the first place. This philosophy emphasizes separateness from nature and control over its processes, as opposed to unity and humility. Eli continued:

> Elders have always taught that they are not worried about Mother nature, they are not worried about the earth, she will evolve and adapt to the new conditions, but we have to get out of her way so that she can evolve and adapt . . . Indigenous Elders see Mother Earth as cleansing and smudging herself; with what people term "natural disasters," Mother is renewing herself with water and fire. Elders advise that we shouldn't fight her but should listen, respect, and work with her.
>
> (Eli Enns 2019)

This Indigenous worldview that sees climate change as a relational issue opposes the Western instrumental value of nature that primarily sees it as an emissions problem. Many people today assert that the threats projected by climate change and pollution can be solved with cutting emissions to zero and implementing engineered solutions. From the Nuu-Chah-Nulth perspective, if the world drastically changes its energy sources from fossil fuel to renewable energy, people may still maintain an exploitative relationship with nature. Behavior can change with the employment of technological solutions, but what is truly required is an internal shift – a change in values. The local Nuu-Chah-Nulth conviction is that climate change is a wake-up call to restore balance and recognize our coevolution with nature.

When Eli asked his elders about the future of *ḥumiis* in response to climate change, they said, "*quay'qwiiksup heshook-ish tsawalk*," which translates to "everything is one and everything is interconnected and everything is in that total state of transformation all the time." Coevolution in the Nuu-Chah-Nulth language was described to Eli by elder Levi Martin; it is captured in *quay'qwiiksup*. Today, in this moment, *quay'qwiiksup* is the temporal equivalent of interconnectivity. "*Heshook-ish tsawalk* means everything that is currently existing today is interconnected. *Quay'qwiiksup* shows how everything that is existing today is a snapshot in an ongoing, ever-transforming reality. It's a cross section through time" (Eli Enns 2019). While ecological relationships between species over space are accepted, Nuu-Chah-Nulth teachings equally recognize the temporal connection as species coevolve over time.

An important part of human-nature coevolution is the reoccurring cycles of birth and death. Eli explained that in his tradition, death is not only accepted and honored but is seen as a necessity. For example, when a *ḥumiis* tree dies, it does not vanish from the world, but as a snag or nurse log it continues to bring life by providing nutrients and habitat for surrounding living species. Nuu-Chah-Nulth individuals actively participate in this wholistic process when they leave reciprocity of some kind upon harvesting parts of the *ḥumiis* (e.g., food scraps) to show their gratitude and respect. Eli described this process further:

> North American [present-day] society does not want to see death, does not want to see garbage. There never used to be such thing as garbage in my territory on the west coast of Vancouver Island. When you had a fish, you had fish bones that would go back into the environment . . . all of the so-called waste products of our economy, of our relationship to the land was part of a natural reciprocity with the land. When you take something from the environment, you leave something, that's what the older people always said. But then when you had the advent of plastics and chemicals, in the 30s, 40s. Prior to that, you had shell middens. How were those created, through people eating clams and leaving the shells through a natural process. There was no such thing as garbage.
>
> (Eli Enns 2019)

Quay'qwiiksup heshook-ish tsawalk teaches the natural cycles of birth and death through its interconnectivity through space and time. Indigenous People would act accordingly based on these values by returning their "waste" back into the environment, feeding new life.

With his acknowledgment of death as a natural and integral part of the coevolutionary process, Eli explained that there may be a time when *ḥumiis* and other historically culturally significant species, like salmon, are no longer existent. He described how all species are on their own personal evolution, many of which provide important coevolutionary roles but can often go unnoticed by humans. As ecosystems and climates shift, various species assume a range of roles. Eli elaborated on this topic:

> There's a totem pole in Tofino and the sun-moon crest, *iisaak*, is at the top and if you were not paying attention, you would not see the smallest totem on that totem pole. It's at the very base; most people don't see it. It's a mouse, a very small carving of a mouse at the base of that totem pole. That mouse doesn't represent just other mice, it represents all of the substructure of organisms and creatures in the environment that are unseen often but it's the foundation of ecological integrity which is more quickly evolving to external influences.
>
> So yes, the cedar tree is important but as our environment changes, it could very well be that various organisms and other species of plants will become more culturally prominent over time. We can't freeze transformation; we have to learn to respect it and move with it. Our inner world first has to be tuned in to the transformation, then our words and our actions flow from our inner worlds. And so the cedar tree represents a chapter in the evolution and transformation of our territories. What I know is that we would want the cedar tree to continue on but as the precipitation patterns change and as the temperature variances change, we can understand that it very well could be that the cedar tree may no longer be able to survive.
>
> (Eli Enns 2019)

Undoubtedly, the extinction of *ḥumiis* would be a woeful story, but a story that would teach and honor it, respecting its part of the whole and the coevolution of all species with time. Oral tradition is a lifeblood for ancestral teachings and an avenue where cultural ecological relationships are maintained and fostered (Atleo 2004; Turner 2014). It is through the passing down of story that cedar, salmon, and others will always stay alive, as through a knowledge transfer where lessons – like the present-day destruction of nature – are shared to teach future generations how to have balanced relationships. The story of *ḥumiis* is already one of immense sanctity and coevolution.

Eli explained that as climates change and ecosystems shift, there may be a new tree of life to take *ḥumiis's* place. Within the sorrow for such a devastating loss, there is resonant faith. Nuu-Chah-Nulth members trust that Mother Earth is not punishing us; she still loves unconditionally despite humanity's destructive behavior. Again, this worldview does not condone environmental destruction. *Quay'qwiiksup heshook-ish tsalwalk* teaches active resilience through difficult changes with trust and an intention to restore and maintain balance that is at the heart of human interconnectedness with the natural world.

> If you have an intact cedar rainforest, it will be able to adapt and evolve in the face of climate change. So we're not turning our back on the cedar; the perspective goes back at minimum 10,000 years within our territory. The transformation stories go back to the time of ice and water and so we've seen, collectively through the generations, great transformation. A lot of the place names that I've been taught talk about that transformation. And yes, it will continue to transform, and as much as possible, you want to continue to maintain a relationship with those highly evolved plant and animal species, such as the cedar. But that doesn't mean we turn our back on the mice and so-called lower forms – those are equally valid.
>
> (Eli Enns 2019)

Joe Martin also emphasized the need to focus on old growth forests, not just old growth trees. The forest is comprised of a highly complex web of relationships, such that the preservation of one species will not support the whole system.

> Sometimes I feel that too much emphasis is put on *ḥumiis* alone, but in fact it's the entire forest. The entire forest was regarded like our church, not just the cedar tree – the entire forest. Of course, cedar played a big part in Indigenous life and culture as did other plants and trees.
>
> (Joe Martin 2019)

Therefore, the cultural and ecological significance of *ḥumiis* for Nuu-Chah-Nulth individuals is defined not just by specific instances of *ḥumiis*-people communion but by the greater socio-ecological network between humans and all life forms. This interwoven network ultimately gives rise to a unique *ḥumiis*-human coevolution.

Meares Island Tribal Park

It may seem that people who view climate change as a moral issue and not one exclusively about emission reduction are focused on a personal debate, and thus are not invested in taking action against environmental destruction. However, there is a difference between defending lands to preserve values as opposed to fighting for only a change in behavior (i.e., switching to renewables) that won't address the underlying cause that is the disconnect between people and nature. Greenhouse gas emissions certainly need to be reduced and renewable energy sources must be employed, but most importantly, humanity is being urged to reconcile its exploitation and superiority over nature and find resolution in solidarity.

The Clayoquot Sound protests on Wah'nah'juss Hilth'hooiss (Meares Island) is one example that demonstrates how Nuu-Chah-Nulth individuals and non-Indigenous allies seeking a shift in human land use ethics were not passive bystanders (Figure 6.1). From 1909 to 1984, Eli Enns explained that his people lost control over their lands and waters. The residential school system grew, and colonists enforced the British North American Act of 1867 that prohibited Indigenous laws, ceremonies, and practices. In the 1950s, industrial forestry production on provincial crown land rose, and forests were increasingly high-graded in BC. Large forestry companies secured permits for large swathes of land, quickly changing the province's landscape within a few decades. One of the most notorious companies was MacMillan Bloedel who obtained a permit to log old growth forest on Meares Island in 1984. However, Tla-o-qui-aht and Ahousaht members along with their allies held a peaceful blockade on Meares Island and declared the Wah'nah'juss Hilth'hooiss Tribal Park in November that year.

Following the establishment of the Meares Island Tribal Park, tensions continued to rise to a peak in 1993 when the BC government introduced the Clayoquot Sound Land Use Plan. The proposal allowed for two-thirds of Clayoquot Sound crown land to be logged. It immediately galvanized a public outcry in response for five months in the summer of 1993. The protest has become known as the "largest act of civil disobedience in Canadian history" where over 11,000 protestors took part in blocking logging trucks, and over 800 were arrested and tried in the BC Supreme Court (Grant 2010). In late 1993, the BC government signed an agreement with the Nuu-Chah-Nulth Nation that allowed for Indigenous review of logging plans while the land dispute was settled. Consequently, the Scientific Panel for Sustainable Forest Practices in Clayoquot Sound was established that was comprised of biologists, Indigenous elders, engineers, and foresters (Clayoquot Scientific Panel 1995). The Panel's members embodied expert knowledge on local forest management, including its regional biodiversity, ecology, and traditional land use separate from government, industry, or environmental organization opinions. At the heart of the Scientific Panel's guiding principles were the interconnectedness of the world at all levels, human respect for nature, and resilience to the inevitability of change long-term, both socially and ecologically (Clayoquot Scientific Panel 1995). Both the Panel's inception and outcomes were landmark events where industrial forestry and government giants conceded to restore ecosystem-based values, launch long-term conservation plans, and conserve Nuu-Chah-Nulth culture in Clayoquot Sound.

The Meares Island protests and Clayoquot Sound Scientific Panel act as examples for modern-day activism efforts and management regimes where Indigenous knowledge systems and Western science can interface. Eli described this intersection as an "ethical space" where Indigenous knowledge and Western science can join in a co-creative space to preserve natural biodiversity. Despite the success in reclaiming Meares Island, both Eli and Joe clarified that the idea of a "park" is fundamentally not supported by an Indigenous worldview. A park implies that what is inside the park is protected and somehow more important than the surrounding, unprotected lands. This idea contrasts *heshook-ish tsalwalk* because it creates separation and fragmentation. Nonetheless, Eli and Joe continue to be committed to reclaiming their ancestral lands and reinstating traditional stewardship and governance. Due to efforts made by Eli, Joe, and their communities, the Nuu-Chah-Nulth nation declared Ha'uukmin (Kennedy Lake) a Tribal Park in 2007 and Tranquil Creek and Esowista (located on Long Beach) Tribal Parks. Innovative clean energy projects, housing, and ecological restoration programs are underway within the Tribal Park areas as Nuu-Chah-Nulth leaders integrate advanced technology and Indigenous wisdom.

Cultural attunement

The respect and nonviolence shown for our peers should not find its limits at humanity. We cannot separate ourselves from our environment and the species we share this planet with, so the principles of how we interact with and relate to nature should be no different than with another human being. Nuu-Chah-Nulth teachings encourage this shift in values – to see and feel *ḥumiis* as a peer and find attunement with nature holistically. For the Nuu-Chah-Nulth people, "climate change is a firm, but loving hand" (Eli Enns 2019). It is a chance for our collective human relationship to shift from dominion to communion.

> Mother nature will heal and recover; we need to stop doing harm. Then we can sit back and see how she is adapting to climate change – how she is healing . . . truly being *iisaak*. By observing mother nature, we can organize our internal worlds better and our words and understandings will flow from that.
>
> (Eli Enns 2019)

To mainstream Western society, climate change is a crisis narrative that requires only change of actions (e.g., bioengineered solutions to greenhouse gas emissions) not change of heart. In the time of significant biocultural loss, Nuu-Chah-Nulth and other Indigenous leaders stand as invaluable teachers for modern society to remember their inextricable connection with the natural world. Through *quay'qwiiksup heshook-ish tsawalk*, they teach people to be respectful and resilient to change. "Paddle with the tide, not against it" (Eli Enns 2019). By trusting the coevolutionary process with nature and being active voices for biocultural preservation, environmental stewardship and Indigenous culture can be valued and restored.

Klecko klecko, "thank you."

References

Atleo, E. R. 2004. *Tsawalk: A Nuu-chah-nulth Worldview*. University of British Columbia Press, Vancouver, BC.

Atleo, E. R. 2011. *Principles of Tsawalk: An Indigenous Approach to Global Crisis*. University of British Columbia Press, Vancouver, BC.

BC Ministry of Environment. 2016. *Indicators of Climate Change for British Columbia, 2016 Update*. Ministry of Environment, British Columbia, Canada. Available at: https://www2.gov.bc.ca/assets/gov/environment/research-monitoring-and-reporting/reporting/envreportbc/archived-reports/climate-change/climatechangeindicators-13sept2016_final.pdf.

Cameron, E. S. 2012. Securing Indigenous Politics: A Critique of the Vulnerability and Adaptation Approach to the Human Dimensions of Climate Change in the Canadian Arctic. *Global Environmental Change* 22: 103–114. https://doi.org/10.1016/j.gloenvcha.2011.11.004.

Case, M. J., and J. J. Lawler. 2017. Integrating Mechanistic and Empirical Model Projections to Assess Climate Impacts on Tree Species Distributions in Northwestern North America. *Global Change Biology* 23: 2005–2015. https://doi.org/10.1111/gcb.13570.

Ceballos, G., P. R. Ehrlich, A. D. Barnosky, A. García, R. M. Pringle, and T. M. Palmer. 2015. Accelerated Modern Human-Induced Species Losses: Entering the Sixth Mass Extinction. *Science Advances* 1: 1–5. https://doi.org/10.1126/sciadv.1400253.

Clayoquot Scientific Panel. 1995. *First Nations' Perspectives of Forest Practices in Clayoquot Sound*. Scientific Panel for Sustainable Forest Practices in Clayoquot Sound. Report 3 (with appendices).

Drake, D. C., R. J. Naiman, and J. S. Bechtold. 2006. Fate of Nitrogen in Riparian Forest Soils and Trees: An 15N Tracer Study Simulating Salmon Decay. *Ecology* 87: 1256–1266. https://doi.org/10.1890/0012-9658(2006)87[1256:FONIRF]2.0.CO;2.

Garibaldi, A., and N. Turner. 2004. Cultural Keystone Species: Implications for Ecological Conservation and Restoration. *Ecology and Society* 9 [online].

Grant, P. 2010. *Clayoquot Sound*. Available at www.thecanadianencyclopedia.ca/en/article/clayoquot-sound.

Huntington, H. P., M. Carey, C. Apok, B. C. Forbes, S. Fox, L. K. Holm, A. Ivanova, J. Jaypoody, G. Noongwook, and F. Stammler. 2019. Climate Change in Context: Putting People First in the Arctic. *Regional Environmental Change* 19: 1217–1223. https://doi.org/10.1007/s10113-019-01478-8

IPCC (Intergovernmental Panel on Climate Change). 2019. Summary for Policymakers. In *Climate Change and Land: An IPCC special report on Climate Change, Desertification, Land Degradation, Sustainable Land Management, Food Security, and Greenhouse Gas Fluxes in Terrestrial Ecosystems*. United Nations, Geneva, Switzerland.

Jakob, M. 2000. The Impacts of Logging on Landslide Activity at Clayoquot Sound, British Columbia. *Catena* 38: 279–300. https://doi.org/10.1016/S0341-8162(99)00078-8.

Liverman, D. M. 2009. Conventions of Climate Change: Constructions of Danger and the Dispossession of the Atmosphere. *Journal of Historical Geography* 35: 279–296. https://doi.org/10.1016/j.jhg.2008.08.008.

McKenney, D. W., J. H. Pedlar, R. B. Rood, and D. Price. 2011. Revisiting Projected Shifts in the Climate Envelopes of North American Trees using Updated General Circulation Models. *Global Change Biology* 17: 2720–2730. https://doi.org/10.1111/j.1365-2486.2011.02413.x.

Meffe, G. K. 1998. *Conservation Biology*: Into the Millennium. *Conservation Biology* 12: 1–3.

Metsaranta, J. M., C. C. Dymond, W. A. Kurz, and D. L. Spittlehouse. 2011. Uncertainty of 21st Century Growing Stocks and GHG Balance of Forests in British Columbia, Canada Resulting from Potential Climate Change Impacts on Ecosystem Processes. *Forest Ecology and Management* 262: 827–837. https://doi.org/10.1016/j.foreco.2011.05.016.

Mobley, C. M., M. Eldridge, C. M. Mobley, and M. Eldridge. 1992. Culturally Modified Trees in the Pacific Northwest. *Arctic Anthropology* 29: 91–110.

Murdock, T. Q., S. W. Taylor, A. Flower, A. Mehlenbacher, A. Montenegro, F. W. Zwiers, R. Alfaro, and D. L. Spittlehouse. 2013. Pest Outbreak Distribution and Forest

Management Impacts in a Changing Climate in British Columbia. *Environmental Science and Policy* 26: 75–89. https://doi.org/10.1016/j.envsci.2012.07.026.

Pojar, J., and A. MacKinnon. 1994. *Plants of Coastal British Columbia: Including Washington, Oregon & Alaska*. Lone Line Publishing, Vancouver, BC.

Reimchen, T. E., D. D. Mathewson, M. D. Hocking, and J. Moran. 2002. Isotopic Evidence for Enrichment of Salmon-derived Nutrients in Vegetation, Soil, and Insects in Riparian Zones in Coastal British Columbia. *American Fisheries Society Symposium*: 59–69.

Stewart, H. 1995. *Cedar: Tree of Life to the Northwest Coast Indians*. Douglas and McIntyre Ltd., Vancouver, BC/Toronto, ON.

Turner, N. J. 2014. *Ancient Pathways, Ancestral Knowledge: Ethnobotany and Ecological Wisdom of Indigenous Peoples of Northwestern North America*. McGill-Queen's University Press, Montreal & Kingston.

Turner, N. J., Y. Ari, F. Berkes, I. Davidson-Hunt, Z. F. Ertug, and A. Miller. 2009. Cultural Management of Living Trees: An International Perspective. *Journal of Ethnobiology*. https://doi.org/10.2993/0278-0771-29.2.237.

Turner, N. J., and H. Clifton. 2009. 'It's so Different Today': Climate Change and Indigenous Lifeways in British Columbia, Canada. *Global Environmental Change* 19: 180–190. https://doi.org/10.1016/j.gloenvcha.2009.01.005.

Väliverronen, E. 1998. Biodiversity and the Power of Metaphor in Environmental Discourse. *Science Studies* 11: 19–34. https://doi.org/10.23987/sts.55111.

Weingart, P. 1995. 'Struggle for Existence': Selection and Retention of a Metaphor. In *Biology as Society, Society as Biology: Metaphors*, edited by S. Maasen, E. Mendelsohn, and P. Weingart, pp. 127–151. Springer, Dordrecht, Boston and London.

Wilms, J., and N. J. Turner. 2014. *Aromatic Oils in Coniferous Trees of the Central Coast of British Columbia*. Quadra Island, British Columbia.

Zahn, M. J., M. I. Palmer, and N. J. Turner. 2018. "Everything We Do, It's Cedar": First Nation and Ecologically-based Forester Land Management Philosophies in Coastal British Columbia. *Journal of Ethnobiology* 38: 314–332. https://doi.org/10.2993/0278-0771-38.2.314.

Part II

Eurasia and Australia

7 Chestnut time and chestnut place

Conserving chestnut-ness (*kestanelik*) in Turkey

Jeffrey Wall

Introduction

Kestanelik translates from the Turkish most roughly as chestnut-ness, but of course, that does not translate it at all. The work of this chapter is to show why this is and to explain how this non-translation looms over humanity's struggle with a kaleidoscopic whorl of environmental danger. Getting to chestnut-ness, hereafter *kestanelik*, will require twin and parallel conceptual tracks. On one side, my chapter's content must bind to actual trees and their relationship with actual, particular people. On the other, it will be necessary to note the flashing implications of this content for the broad practice and rhetoric of environmental conservation, way beyond these trees and these people. The evidence I will present for this task will be a weave of ethnographic observations – made during months of fieldwork across Turkey between 2015 and 2017 – and more flowing and personal narratives of learning, insight and discovery, which I experienced as I sought understanding of the value and esteem for the chestnut species in Turkey, not just in chestnuts and chestnut trees, but around *kestanelik*.

Scenes and observations

A very particular scene can give myself and readers good initial footing. This scene – one in fact captured on film – holds a story underneath it. First the scene: in a mature *kestanelik* – read chestnut grove – a large loose stump is being slowly devoured by a bed of coals and light licking flames. Around this hearth, a barrel-chested, mustached man scatters a ring of small, Black Sea chestnuts, known affectionately as *kuzu kestane* ('lamb chestnuts'). Like rings around Saturn, this belt of chestnuts encircles the center, the heated, glowing coalbed topped by the massive, upturned stump. Enter the man again, this time with big armfuls of wilted ferns, piling them over the chestnut ring, yet saving a more verdant bunch of them to grip tightly by the stems, thereby making them into a fire tending instrument like a combined fan and brush. With this impromptu tool, he dusts and dabs at the ferns, deftly coaxing heat and then flame onto the mounded fern loop atop the chestnut ring.

DOI: 10.4324/9780429320897-9

Sounds are what next dominate the scene. The crackle of the ferns receiving flame, followed by a thud of sudden gas and pressure release beneath the upward bursting flames. At this point, what has been consistent chirping and banter of an off-camera audience crescendos into oohs and aahs, guffaws of laughter and excited speech. Flames rage no more than fifteen seconds all around the hearth, enveloping all including the whole stump. Then they die and the man attacks the remaining ashes suddenly with the fern fan. The bursts of air he causes dissipate the debris, soot and ash instantaneously, revealing a scorched belt of earth dotted with shining auburn chestnuts. A delicacy revealed is met with hushed awe of the off-camera audience, allowing in the gentle soundscape of a forest. End scene.

Figure 7.1 Central stump fire, public excitement, *kestanelik*, Abdülkadirköy.

I choose to communicate the origins of this scene by explaining my presence there. It is a social, even personal story which only later bore 'scientific' fruit, and even that, as I'll explain, required significant further processing. The beating heart of our investigation of value for the chestnut species across Turkey was the question: "who cares about chestnuts?" The meaning of that question for us was open to every possible interpretation, from the pejorative sense, as in "who cares about chestnuts anyways?," to specific names of people. After all, much like in North America, the decline and endangerment of the species was well-known in the Turkish public, and to many it amounted to just another written-off environmental and cultural loss in an era full of such things. That *kestaneler kuruyor* ('chestnuts are drying up') was common knowledge, from the city streets to the highland villages. So, within the ranks of countless storylines of loss which weave together most people's experience of modernity, who still cared about the chestnut was no simple question to answer.

Jamil did care. The man in the scene. The man with the ferns. Months before the scene, he participated happily in our initial questionnaire. In doing so, he not only answered any and all questions thoroughly and instantaneously, he also suggested more questions which we should have asked, more morphological observations than our questionnaire was designed to elicit. Going further, he told us exactly what we should tell forest bureaucrats and he named names: names of officials, names of offices. He had made campaigns of various sorts having to do with the local chestnut population over the years, the mental maps of these clearly remaining very fresh in his mind. In short, talk of the tree and thought of the tree animated him beyond the usual bounds. As part of my research program, I sought to see more of what he saw, to catch more of his contagion. As it turned out, this was a great first step in getting to *kestanelik*.

I asked him and my colleagues if it would not be a good idea for me to rejoin he and his family in their village of Abdülkadirköy during the harvest period. He loved the idea, but clearly took this talk on my part to be idle pleasantry, rather than commitment. In any case, his enthusiasm was again palpable over the phone when I reached out to him the following September, my intentionality and sustained interest apparently serving as an important fuel for trust. I was to come up for daily work for a week, collecting chestnuts for his household.

Seeing more purpose in livelihood

By way of setting the stage for the previously described scene, my first entry in my notes on day one of my harvest visit reads precisely like a cast of characters. In this case, these became the off-scene spectators, the ones oohing and aahing:

> On this day the following characters were seen and engaged with at the kestanelik – read chestnut grove: Samir, wife of Samir, Son of Samir (Ali), a delikanli (a 'crazy blooded' or young man) and his abla (older sister), a trio who live in Holland (relatives? father + mother + daughter), one lone

> female age around 40, an avocat (lawyer) who brought one fish, a city-fied older couple, a village muhtar (elected representative) from another village.

While my later shifts in this *kestanelik* did reveal its quieter, more meditative times, the insight still stands firm from day one that this remote forest location is seen, presented and experienced as highly public. As such, the infrastructure of the scene was apparent: the impressive, primordial hearth, an adjacent sprawling tarp covering 5–6 square meters, the *söfre*, or dining area/mat, tea pots, a pile of work implements such as gloves, tongs, *ünlü* (aprons). In short, the scene I described previously only boosted my sense of understanding once I grasped the need to expand an interpretation of *kestanelik* from the place of chestnut collection – and later, use – to include the time spent engaged within it, including its public and personal dimensions.

Complementing my learning of this highly public, performative role for *kestanelik* was an emerging grasp of the lived essence of this livelihood tradition which participants conveyed to me in Abdüldkadirköy and other sites, an essence that was at variance with conventional understandings of the term, livelihood, clouded as they tend to be with utilitarianism and notions of brute survival. It was my study in the Caucasus (Wall 2012) which opened my eyes to the broader *telos* or purpose of chestnut livelihoods to cultural continuity. Stated briefly, to the cultural minority Avar and Tzakhour People, inordinate value was placed on their own numbers and presence in their homeland. This was realized in the situation I came to observe, where, at the level of community, harvesting and sale of chestnuts from traditional commons was encouraged for families raising children. My time spent collecting chestnuts across Turkey's geography at variable economic intensities further broadened my perception of the spectrum of value and importance attached to a particular traditional ecological management and subsistence practice. The economics of chestnut, I discovered, were best understood as harnessed, to a major extent, to the larger value of cultural continuity and place fidelity.

This keeping-culture purpose of chestnut livelihoods revealed itself in more and more kaleidoscopic patterns the more I engaged it as a researcher. There was a man I tailed for a full day who inspirited his day of harvest labor with the thought of using these *düşen paralar* ('falling coins') raining down from the forest to surprise his daughter with one month's rent for her Istanbul apartment. Other thoughts animated him as well though, and these were revealed like layers of an onion. He explained how he revered his father's image as an active man and that he loved conjuring this image through his own labors to collect chestnuts. He articulated these motivations before showcasing another: a half hour break to watch and laugh at his dog playing in the creek. In this vein, I learned during the quieter days at Abdülkadirköy that Samir's three sisters, Reyhan, Jamila and İpek, did not in fact live nearby, nor did they need the money from chestnut sales. The truth appeared to be that they leveraged the classical obligations of family labor (Morvaridi 1992) to broker this time away from their husbands and children in order to see each other, if only for a day or two. More, they chose to see each other in this specific context so rich with special childhood memory. They said they did not convene routinely as sisters at any other time of year. At this time and place (*kestanelik*), they dwelt in thoughts of their mother, father and their upbringing. So faint was the economic logic in scenes of delight like these that it seemed rather forced.

Grandfather popped corn, grandma roasted chestnuts (*Dedem mısır pattlatı, nenem kestane pışirdi*)

I would recount scenes from my multisite ethnography – like that from Samir's *kestanelik* – everywhere I could. I would describe them and/or show videos on my phone to get a sense of how they were received with the various individuals I spent time with over the years. In the same vein as the fern-fired chestnuts, I could add stories of men drinking towards dusk and firing antique pistols, or a family hunting down and killing a snake in the orchard, holding its dangling body high overhead like a trophy. Such tales came as little surprise to most recipients, or more accurately, they folded easily into a well-worn genre of *açık havada eğlence* ('outdoor fun and amusement'). This can be understood broadly to include the hilarity, shenanigans, bravado and tomfoolery – of men, women, children, grandparents, visitors and, of course, *delikanlı* ('crazy bloodeds,' i.e., young adult men) alike – when excited by being together in the outdoors. *Kestanelik* were viable settings for this well-grasped cultural excitement as were *zeytinlik* ('olive groves'), *ormanlar* ('forests in general'), *yayla* ('highland pasture') and many more.

However, another genre of experience from *kestanelik*, during my aforementioned process of reporting and recounting, consistently rung more of a bell with my conversational partners. The scene I joined and disseminated which hit this note goes as follows: Samir invited me and my family to his home to have dinner on one night of the harvest period. His wife, Samina and two sisters were present. Jamila was younger than he, and İpek older. As mentioned, they were both at present visiting their brother, who resided in and maintained the home where they were raised. Upon arriving, we were welcomed at the car and invited upstairs to a cozy living and dining room. The slight chill outside meant the burning wood stove was a welcome presence to all. Conversation was a free-for-all between the genders and ages, with my wife fielding questions from Samir and me translating a fun inquisition of my older son by Samina, Jamila and İpek. As per custom, Samina was zipping around conjuring a soon-to-arrive meal. Dinner was simple pasta, simple to serve and clean. My wife's and my efforts to oversee our young children's eating were a source of something like dismay, combined with empathic warmth and comedic effect.

In what moments I could, I scanned for *kestanelik* around me. I noted chestnut framing around the windows and doors. The floorboards showing between the several rugs around the room were likely chestnut. These uses were extremely common, in both the Black Sea and Aegean regions where chestnut trees grew. I understood their common-ness and persistence as important in their own right. Still, when I asked to verify these chestnut materials, the answer, as usual, was a humdrum, "Yes, that's right. Chestnut."

Chestnuts were on the menu after dinner and desert, which was cake. Jamila, the younger sister, in a lightly reclined position at the dining table, held a bag of that day's collected chestnuts, using a sharp paring knife to score each with an 'X' before laying them gently on the wood stovetop, scored-side-down. Some chatter about these techniques was mildly enjoyable to Jamila and İpek. Samir joined in to add details. Still, the kind of smiles I was seeing and the kind of enjoyment I detected read more to me like, "isn't this just funny, us fielding these curious questions from this curious foreign guest?!" Then something dawned on me. While the TV was on, it was barely audible, and no one was really paying attention to it. This struck me as ever so rare. My guesting skills in the larger region, developed over decades at this point, from the Caucasus to Egypt to here, had hardened me to pursue meaningful conversation in the face of loud and flashing television entertainment. It was and is an undeniable hardship for me, and one which doubtless gave rise to my strong preference for outdoor interviewing about outdoor subjects. As a very brief aside, this habit I only later understood had had serious gender implications in my work, as women in the region were by and large more comfortable with interviewing indoors (Wall et al. 2018).

In any case, these thoughts were occurring to me during this very unordinary way of passing time. The slowly arising sounds and smells of roasting chestnuts in the room, the warmth and brittle crunch of the nut shells on our fingers and their gratifying removal – a beloved attribute of these *kuzu kestane*, ('lamb chestnuts') – the idle chatter between us, so rich in comparison to the usual tinny and grating sounds of the television: all of this fed into my heightened awareness of the unique quality of these moments. I suddenly had to speak this understanding, had to validate it. In a quieter moment, I asked, "*bu zaman da kestanelik olabilir mi*?" ('Can this time also be considered kestanelik?') The question rung out then, just as it has rung out in my retellings of this scene. "*Tabii ki canım!*" ('Of course, my soul!') came the reply from İpek, "*zaman kendisi kestanelik olabilir, ne güzel koymuşsun*!" ("time itself can be kestanelik, how beautifully you have put it!")

This reaction, in spirit, has been reliably elicited from a number of audiences. Whether a room full of serious forest bureaucrats, Airbnb hosts in Istanbul, cab drivers in Ankara, all of these have similarly appreciated or been struck by this poetic twist toward understanding. The chair of Hacettepe University's Department of Anthropology at the time, Dr. Yilmaz Erdal, took this account in and nodded saying unprompted, "*dedem mısır pattlatı nenem kestane pişirdi*," ('grandpa popped corn, and grandma cooked chestnuts'). He explained that this was a figure of speech, and one he knew deeply as both a local cultural studies enthusiast and as a boy from a forest village of eastern Anatolia. Its literal meaning is enveloped in its collective recognition as a harkening to home, history, homeland, kin and destiny, as an archetypal and world-making invocation. Reference to and recognition of this invocation, with or without chestnut, remains ubiquitous in contemporary culture. As child artist İrem Baltacı very recently declared in a press interview, "*Hayatın anlamı dedemin patlattığı mısırı yiyip ailecek hep beraber sohbet edebilmektir*" ('The meaning of life is to eat the popcorn my grandfather made and to make conversation together with family') (Hakyemez 2021). The significance of the presence of chestnut species within this invocation, though regionally specific, will be discussed more fully in the next section. In preparation for entering such a symbolic register, it makes sense to first deal with the more immediate poignancy of chestnut's tactile physicality and its subsequent ontological irreplaceability, since, at the end of the day, our immediate task is really to size up the lingering care and affinity for this specific, threatened species.

Harnessing aesthetic motivations in conservation thinking and practice

Our investigation into who cares about chestnut was always an avenue toward probing the underexplored textures and fabric of value for the species at a notable moment when its population, health and productivity were in precipitous decline. So far, just a few scenes or ethnographic encounters have been magnified in any detail, but they do contain sufficient context to suggest a few important themes. I would argue that the first is a flashing indication to revisit the power of aesthetics to inspire conservation action through some combination of meaning and affinity. Trivial though it might seem, in the story of my own understanding, this flare was lit and launched at the moment I saw that these aesthetic properties were powerful enough to light a path away from television. The doubtless fact that other active ingredients were involved in this force will be discussed in due course, but it remains the case that chestnut, and not the fruit of another life form, was central to the ceremony I observed. This takeaway is easy to augment from within my bank of observations of appreciation for countless other raw material expressions of the plant. There was the delight explained by boat builders of feeling and trusting the plasticity of chestnut boards (specifically those from northeast Turkey) when they are strapped onto the curvature of fishing ship bows. There was the nearly universal enlivening and enveloping caress of roasting chestnut scents during the fall for millions of Turks walking the bazaars and boulevards. Many remarked on the incandescent flowering, bursting like fireworks in the tree canopy, enjoyed by diners, walkers, drivers, flight and ferry passengers alike. All of these illustrations, seen from a certain angle, sketch out a powerful bond between the inimitable physical expression of the chestnut tree and vast active living cultural systems.

Along with overwhelming evidence of the chestnut tree as an aesthetic powerhouse, participant testimony and general observations expose an ontological value engrained within this appreciation. This occurs primarily through declaration and recognition of its inimitable and irreplaceable place in life, a status which cross-cuts gender (Wall et al. 2018), status, ethnicity, livelihood and more (Wall et al. 2019). When envisioning a total decline or disappearance of the species, participants dive philosophical with exclamations such as, "*kestane olmasa nereye dönelim?*" ('If there was no chestnut, where should we turn?'). Another man claimed that he went into the chestnut candy business because daily cognizance of these trees kept him cognizant of "*nereden geldik, nereye gideceğiz*" ('where we came from and where we are going'). A young woman sighed mournfully, "*kestane yaşamda ve ölümde bizim evimiz*" ('chestnut is our home is life and death'). In other words, considerations of its endangered and declining status inevitably led to an emotional register of loss and insecurity. This consistent turn, though reached via the engagement of the trees' aesthetic properties and their value, eventually steered our gaze down a deeper hallway.

It remains the case that tactile appreciation, olfactory delight and visual stimulation, even if bonded to an ontological mandate for the chestnut species, fall short of the goals of this chapter in two related ways. First, this field of motivation, substantial though it may seem to many, is not known to rise in the modern period to a level of consequence necessary to supersede the conflicting force field of utilitarianism. As Passmore states bluntly, "any argument for preservation founded in utilitarian instrumentalism will allow that economic considerations, in a broad sense of the phrase, might under circumstances outweigh the case for preservation" (Passmore 1974: 25). It should come as no surprise that in the case of chestnut decline across North America and much of Europe, economics, and utilitarian forestry more broadly, has not judged the chestnut species to be exactly irreplaceable or even worthy of necessary efforts for protection and maintenance. Therefore, in order to envision a plausible future for endangered and beloved species like chestnut in Turkey at a moment where utilitarian valuation generally reigns, it will be necessary to identify, understand and support motivations profound and enduring enough to feasibly supersede utilitarianism. Fortunately, there is a second sense in which aesthetics does not meet the moment, and one which points the way to a force which might.

Chestnuts and enchantment

In all the scenes I have sketched in this chapter and in most of the aesthetic joys observed and declared, there lurks the radiation, the throb, of more terminal (à la Rokeach 1973) or intrinsic values. Despite their palpable presence, these had a way of remaining unnamed in participant reporting and in the inner monologue of my ethnographic observation. The most ready and obvious reasons for their elusive and concealed quality are as follows. First, most participants felt and demarked a concrete and formidable barrier between the implicit register of formal interviewing and the inner family/community sanctum in which these values were harbored and fed. To illustrate, I had a technique – in the end not a great one – of following up with any participant who claimed they collected chestnuts or other materials from the tree for money. I would counter, "well then why do you collect money?" This inevitably provoked some spike of discomfort for my participants of one sort or another and never once provoked a thoughtful exposition on deeper values. Some would declare in effect, "why to live of course!" or "what kind of question is that?" Others would look pleadingly at my Turkish colleagues as if to say, "could you please step in here and help this man save face?" In any case, the bare chemistry of research behavior, on all of our parts, and with a foreigner in the mix no less, proved perfectly incompatible with the mood or ambiance conducive for such heart-to-heart discussion. This is one reason why harvest visits, when most time was spent outside of dialogue and no formal structure was supplanted onto talking, proved so vital. Yet even in this practice, most acted and realized values were left to be inferred rather than declared directly.

The second obvious reason for the persistently unnamed and concealed status of more deeply operating values – and doubtless a source of strength behind the barrier described previously – is a ubiquitous modern disinclination to pair in the same thought or sentence the quintessentially material physical chestnut tree and its parts with precious abstractions like family love, universal meaning and the like. This discrimination would certainly be operating behind the speech served up by participants catering to the intuited needs of outsider researchers, as well as within my frame of observations.

Getting to *kestanelik*, for me, meant coming to the conclusion that aesthetic appreciation for chestnut, though profound and enduring in its own right, while exceeding economic value by miles, is still cast in a supporting role to a more ultimate social gravity, what I have identified as enchantment. In the sense I have had cause to use it, the definition of enchantment merrily dances between the two presently accepted poles of its dictionary meanings of first, the state of being spellbound, or placed under a spell, and second, a state of being exceedingly pleased and delighted (Dictionary.com 2021). More robust and applicable meaning flows from the etymology and nature of the word to something like *being imbued with song from without*, a meaning more akin to inspiration (Etymology Online 2021).

The raw challenge of bringing enchantment into a public scientific document is perhaps best articulated by philosopher of modernity, Charles Taylor, who believes that for true moderns "thoughts and meanings are only in minds," (Taylor 2018: 35). But as *kestanelik* was agreed by all to be in the very time and space we spent together, as well as in our very togetherness, it well appears that these very gathered and cracked nuts released more than their tender, warm and sweet flesh, they released a meaning received by all, a meaning from out of mind. Herein lies the friction in the fact of enchantment being so well-known, so well-worn, yet maladapted to modern self-understanding. The fundamental first step to understanding this recurring friction is to revisit Max Weber's 'tale' of disenchantment (McCarraher 2015: 86). In this account, the advent of scientific certainty, Protestant austerity, industrialism, free market paradigms and much else stripped the operating worldview of moderns of any notion of a world inhabited by agents and agency, most commonly understood as spirits, moral forces and the like. In Marx's haunting characterization, "capitalism has drowned the most heavenly ecstasies of religious fervor, chivalrous enthusiasm, or Philistine sentimentalism, in the icy waters of egotistical calculation" (Marx and Engels 1848: 16).

As a cluster of philosophers have endeavored to show over the last few decades, however, the clench of this tale of mass disenchantment as a *fait accompli* is surprisingly easy to relax with the briefest glance around one's own local community. Charles Taylor discerns between two core meanings of enchantment, one which he concedes is ascendent, and one which he does not. In the first sense of enchantment as a perception of the world as pregnant with angels, ghosts and magic, moderns are arguably well disenchanted. However, in a second, more elusive sense of enchantment as understanding the world and time as holding out the promise of meaning and purpose, enchantment may well remain for a great many of the citizens of modernity. While the first is easily grasped as contrary to a modern worldview, the second is less obviously so, or is perhaps more recalcitrant. What is unclear, is how it is surviving.

Biocultural liturgies

The relegation of the television to background noise in Samir's house that night will forever stand out as a signpost attracting my thinking to a certain chasm between worlds, to the briefest ascendancy of a deeply familiar yet chronically muted frequency of social life. What is more is that the high physicality of hot chestnuts, warm wooden room and comfortable chairs, was itself the undeniable energy that buoyed said frequency into ascendency. This matter-catalyzed-experience-of-minds is likely a well-observed and well-known fact in individual life histories and intimate kin and community settings around the world. Still, it cuts against what is thought to be the prevailing modern paradigm in a fundamental way.

From my time in Cairo, Egypt, I learned a specific term for matter-catalyzed-experience-of-minds: *mazag*. In colloquial Egyptian Arabic, *mazag* means most literally blend, but is used as a way to more broadly describe some favorable combination of activity, mood, vibe and milieu. I learned the term like any other, by hearing and then using it, but similar to curse words, I picked up instantaneously that it was a word of separate order. It implied mischief in that it could refer to drug highs from hashish and others. It had that property where its very utterance could instantly transition conversational nature and could heighten delight in group settings with its latent content of humor and scandal. Yet, to a new language learner like myself, its highest and most rewarded expression was much more of an art, an incantation of some form of majesty. "Its underlying principle [is] the ritualistic offering of oneself and others of a moment of considered time" (Golia 2004: 169). It was,

> very much a part of a more leisurely pace of life when people could take time to do certain things, however trivial, in a certain way . . . *Mazag* activities have value and meaning in and of themselves, irrespective of any specific result.
>
> (American University of Cairo Professor Mahmoud El Lozy in Golia 2004: 170)

El Lozy's articulation of *mazag* maps well onto what McCarraher calls sacramentality (McCarraher 2015: 93) and Smith calls cultural liturgies (Smith 2009). What renders both these thinkers an asset to the case for *kestanelik* is their compelling portrayal of the enduring power and ubiquity of enchantment in contemporary and even secular life, concealed and muted though it might seem. For McCarraher, modernity is nothing short of a mass mystical cult where capitalism itself acts as a "covert form of enchantment, all the more beguiling on account of its apparent profanity" (McCarraher 2015: 89). What Smith adds to the thesis is the mechanism of the cultural liturgy, a name for the idea "that what we do teaches us how to love" (Gilger 2018). For Smith, this is positively meant to include mundane and repetitive tasks, habits and rituals, from shopping malls and televised sports to sweat lodges. In combination, all of these ideas amount to a surprisingly alternative theory that what simple things we do, *and what we do them with*, indeed build into our loftier meanings and senses of purpose. Put onto *kestanelik*, this can tether small and seasonal nuts to core motivations for life. However, it plays to the strength of this theory that the expression of chestnut trees happens to be brought by human effort into a much more enveloping role than mere snacking on nuts.

Dwelling as an anti-modern

Much as hides and horns make for delightful village and home decorations for a Nilotic people surrounded by living hides and horns, chestnut materials are gathered and placed nearly everywhere around a home in the Black Sea region. In a single household (Figure 7.2), chestnut might very well be the framing,

Figure 7.2 Iconic chestnut house and inhabitant.

floors, windows, doors, siding, stairs, roofing, railing and paneling in homes; the baskets for agricultural labor; and the entirety of an outdoor tea-porch including the benches. Chestnut might be the posts and rails of fencing and the stakes in the garden. This same house, without a shadow of a doubt, roasts chestnuts during many a cool evening. Chestnut floorboards stain the house socks of those who scuttle around within it. A chestnut leaf may be placed beneath their homemade bread, leaving its unmistakable imprint under the finished product. For all of these expressions and more, the durability of chestnut products is cited as a reason for its use and a virtue unto the world.

Much like Tim Ingold found belonging and dwelling as inherent to cultural evolution and persistence (Ingold 2000), there is good cause to survey threats to enchantment with life with Heidegger's understanding of homelessness (*Heimatlosigkeit*). It was precisely Heidegger's intent to resolve 'homelessness' of modern *Homo sapiens* that set him upon his legendary fixation on the meaning of being itself, a fixation marked upon his tome, Being and Time (*Sein und Zeit*). Abstaining from a substantial detour, it will suffice to say that it was Heidegger's determination that all of western thought and philosophy dating back to at least Aristotle had neglected to reckon with the meaning of being and had consequently laid down the bridge into the abyss of homelessness in which we moderns now lie and float and drift aimlessly and frightfully. In his own words, he saw his larger intellectual project as intended to serve humanity's return to "poetically dwelling" on earth (Heidegger 1971: 213).

It would not come as a surprise to many that the philosopher Heidegger was long on theory and short on prescription. However, no biography of Heidegger would dare exclude his fidelity to and his affinity for his native lands of Germany's Black Forest. Nor would any biography of merit exclude his reverence and meditation upon local folkways, practices and pastimes which he consistently cited to illustrate his ideas (see 'Building, Dwelling, Thinking' in Heidegger 1971). Therefore, it is not wayward in the least to cite İrem Baltacı's expression on the meaning of life, or Samir's family's savoring the idea of chestnut time as glimpses of a more poetic dwelling.

Conclusion

What is the value of chestnuts smelling heavenly, peeling effortlessly, being bitten through buttery and as smooth as Turkish delight, and with a slightly sweet and salty aftertaste? What does the knowledge of chestnut longevity mean to a homeowner living inside it? What does it mean to know that the dark meandering grains in chestnut wood around you make it *özel ve güzel* (special and beautiful)? When İrem Baltacı declares that "the meaning of life is to eat the popcorn my grandfather made and to make conversation together with family," and when a good many citizens of Turkey agree whole-heartedly with the proposition, and when a crescendo of chestnut value and meaning arises from my collected observations across Turkey, the role of enchantment comes to the fore. This is because the very structure of İrem's proposition about the meaning of life, so welcomed by common sense morality in its context, envisions ultimate meaning as entirely out of minds, as something that is implicitly shared rather than 'had' and, even more, as something definitively dependent on robust attachment to the material world.

This study goes some way toward showing the necessity of grasping the power of what might be called abstractions within local motivations for environmental conservation. The introduced concepts of *mazag*, sacramentality and biocultural liturgies, which perhaps dovetail with other well-known concepts like *terroir*, are needed to shed light on the hybridity of what makes the bonding agent in biocultural associations. The variations of chestnut morphological expressions are indeed irreplaceable and inherently worthy of conservation care and effort, but if probed with locals, this value is absolutely not to be understood or explained detached from the larger firmament of more abstract meaning and significance in which it all belongs.

I hope to offer the reader a chance, like I have had, to reimagine the gravity and orientation these scenes have offered their participants in their own affective landscapes, life histories and in their perpetual reciprocity of purpose between themselves and their surroundings, their close community and their families. Because I believe it will be especially helpful and productive, I introduced this content to a larger, disparate project of forlorn modern philosophers who have struggled to drive back what they see as an incessant meaningless of modernity by reconciling humanity with earth, time and material. In the final analysis, I believe that the encompassing meaning and enchantment from chestnut trees in Turkey flows onto and submerges their caretakers and offers a vivid example of drive and motive for humanity's essential reconciliation with the more-than-human-living world.

References

Dictionary.com. 2021. *Enchantment*. Available at www.dictionary.com/browse/enchantment.

Etymology Online. 2021. *Enchantment*. Available at www.etymonline.com/word/enchantment.

Gilger, P. 2018. James K. A. Smith's Theological Journey. *America Magazine: The Jesuit Review of Faith and Culture*.

Golia, M. 2004. *Cairo: City of Sand*. Reaktion Books, London, UK.

Hakyemez, A. G. 2021, August 12. İrem Baltacı: "Hayatın anlamı dedemin patlattığı mısırı yiyip ailecek hep beraber sohbet edebilmektir". *Gazete Sanat Çocuk*.

Heidegger, M. 1971. *Poetry, Language, Thought*. Harper and Row, New York.

Ingold, T. 2000. *The Perception of the Environment*. Routledge, New York.

Marx, K., and F. Engels. 1848. Manifesto of the Communist Party. *Marx Today: Selected Works and Recent Debates*: 1–263. https://doi.org/10.1057/9780230117457.

McCarraher, E. 2015. We Have Never Been Disenchanted. *The Hedgehog Review* 17: 86–101.

Morvaridi, B. 1992. Gender Relations in Agriculture: Women in Turkey. *Economic Development and Cultural Change* 40: 567. https://doi.org/10.1086/451961.

Passmore, J. 1974. *Man's Responsibility for Nature*. Duckworth, London.

Rokeach, M. 1973. *The Nature of Human Values*. Free Press, New York.

Smith, J. 2009. *Desiring the Kingdom: Worship, Worldview, and Cultural Formation*. Baker Academic, Grand Rapids, MI.

Taylor, C. 2018. *A Secular Age*. Belknap Press, London, UK.

Wall, J. R. 2012. *The Important Role of Chestnut Production in Northern Azerbaijan and Strategies for Conservation in the Face of Blight Cryophonectria Parasitica*. Cornell University, Ithaca, NY.

Wall, J. R., E. Aksoy, N. Köse, T. Okan, and C. Köse. 2018. What Women Know that Men Do Not about Chestnut Trees in Turkey: A Method of Hearing Muted Knowledge. *Journal of Ethnobiology* 38: 139–155.

Wall, J. R., T. Okan, C. Köse, N. Köse, and E. B. Aksoy. 2019. Folk Biological Value and Chestnut Conservation in Turkey. *Economic Botany* 73: 461–476. https://doi.org/10.1007/s12231-019-09476-x.

8 Hawthorn in Ireland

A rich heritage of folklore, fact and fantasy

Peter S. Wyse Jackson

Introduction

The hawthorn tree, *Crataegus monogyna*, is not the largest tree in Ireland, nor the most beautiful, nor the most useful, but nevertheless it occupies a unique and extremely important part in the cultural history and environmental and agricultural life in Ireland, from the earliest times of human settlement to the present day. If one was to seek to file information on the hawthorn in Ireland under various important categories, if would fit very well under a range of diverse headings, including folklore, mythology, landscape, agriculture, food, medicine, literature and much more.

I am fortunate to have a small property on the Dingle Peninsula in Co. Kerry in southwest Ireland. There, in my 4–5 acres of rough marshy land, are many hawthorn trees, most of them along the old field boundaries where they were probably originally planted as hedges. They have now grown to be substantial specimens with trunks of impressive girth and no longer providing a stock-proof barrier for any animals that wish to pass from one field to the next. Some years ago, a mighty storm swept across the peninsula one night, and the following morning, I was sad to see my largest specimen lying on its side, thrown by the wind. Too large to try lifting it up again, I left it where it fell, unsure what to do. Like most Irish, perhaps I have a natural superstitious caution against cutting down a hawthorn tree. It rewarded my patience, and the following spring, it burst forth into leaf again and ignored my neglect with a spectacular show of blossoms and a heavy fruit crop later in the year. Over the years since then, it has settled into its new horizontal life and grows happily, promising me, and hopefully several generations of my successors, a resilient venerable tree to admire, and bountiful crops of haws each autumn for the birds and me to share.

Hawthorn is one of the most important small trees in Ireland and is found abundantly throughout the country in hedges, fields, scrub, waste places, along riverbanks, woodland edges and rocky places, even growing on walls where birds may have left its seeds. When it occurs in native woodlands, it is an understorey tree, but is more common at woodland edges. Hawthorn is particularly common in lowland areas but can reach altitudes of 1,000 feet (c.300 m) or more in sheltered locations. Indeed, hawthorn is regarded as being the

DOI: 10.4324/9780429320897-10

Figure 8.1 Hawthorn fallen and laying.

commonest tree in Ireland today. Some exceptional specimens may be found, and the current girth champion in Ireland is a specimen measured at 2.65 m in girth (8.69 feet) and 12 m tall (39.4 feet) at Springfield House, Bennekerry, Co. Carlow (Anonymous 2005). Nathaniel Colgan in his *Flora of Co. Dublin* (1904) writes about famous old hawthorn groves of Dublin's Phoenix Park containing many fine trees up to 7 feet in girth. On the Internet, I have come across several references to hawthorn trees living up to 400 years old, but the source of this information is unknown to me. *Crataegus monogyna* is a shrub or small tree, a member of the Rosaceae family, typically 5 to 14 m tall (15–45 feet). It has a dense crown and dull brown bark with vertical orange-coloured cracks. It is very varied in growth form and can occur from an erect tree to a recumbent shrub.

The younger stems have lots of sharp thorns, about 12.5 mm long. Its leaves are variable in shape, 20–40 mm (1–1.5 inches) long, obovate and deeply lobed. The upper leaf surfaces are dark green and paler green underneath. The flowers contain male and female organs (hermaphrodite) and appear in late spring, during May and early June. They occur in corymbs of 5 to 25 flowers clustered together, each flower about 10 mm in diameter, with five white (occasionally pink) petals and numerous red stamens. Each flower has a single style. Bees, midges and other small insects pollinate the flowers. In late summer the flowers are followed by fruits, the haws, which ripen as small oval dark red fruits, about 10 mm long. They are berry-like but have the structure of a pome (a fruit consisting of a fleshy enlarged receptacle and a central core containing the seeds, e.g., an apple or pear). *Crataegus monogyna* haws contain only a single seed. These fruits are an important source of food for wildlife, especially birds such as thrushes and their relatives, and the seeds are dispersed in their droppings. The flowering of the hawthorn is one of the great spring sights in Ireland. The dense heads of the flowers are heavily scented and appear just a few weeks after the leaves appear.

In Ireland, *Crataegus monogyna* is the only native member of its genus *Crataegus*, although several species have been introduced and grow wild or in cultivation in Ireland, mainly for ornamental purposes. Over the years, the common hawthorn or whitethorn has been called various scientific names, including *Crataegus oxyacantha* which was formerly recognized as a separate species and *Mespilus oxyacantha*. A second wild species, *C. laevigata* (formerly called *C. oxycanthoides*) is also found but is probably not native. That species is rare except in the north-east of Ireland where it is frequent. It can be distinguished from *C. monogyna* by having shallowly lobed leaves (unlike *C. monogyna*, where they are deeply lobed). Also, *C. monogyna* has the base of its leaves with pointed lobes, whereas in *C. laevigata* they are blunt. Hybrids between the two are found, relatively frequently in the northeast, and perhaps often planted. Hackney (1992) mentions that the hybrid occurs in some quantity in hedges in the northeast, most of which date from the second half of the 19th century. He suggests that they were probably planted using quicks imported from nurseries in the southern half of England, rather than being grown from local stock. He also highlights recent hawthorn plantings along the M1 motorway, from imported Dutch stock of either *C. laevigata* or the hybrid. There is little or no folklore recorded about any hawthorn species in Ireland, other than for *C. monogyna*.

The name, *Crataegus*, is a Linnean name, published by the great Swedish botanist, Carl Linnaeus in 1753 in his work *Species Plantarum*. It is derived from the Greek word 'kratos' meaning "strength", because of the durability of the wood and 'akis' meaning "sharp", referring to the thorns that are found on some species. The name 'haw', is originally an Old English term for hedge (from the Anglo-Saxon term *haguthorn*, "a fence with thorns") (according to Wikipedia) and the name is also applied to the fruits.

In Ireland, *Crataegus monogyna* is generally called "hawthorn" or "whitethorn", and occasionally "May bush" too. Other English names that have also been used in Ireland include Bread-and-cheese, Drean bawn, Skeeogc, Skayug, Skaig; and in Irish, *Sceach gheal*, *Huath* or *Uath*, *Coirr-sceach*, *Crann sceiche* and *Scé* (Wyse Jackson 2014). The most commonly used Irish name, '*Sceach gheal*' means a "bright thorn". Geoffrey Grigson (1975) in *The Englishman's Flora* gives a long list of vernacular names for hawthorn, including some from Ireland – Mahaw, Quick, Scrog, Scrog-bush, Shiggy, Skeeog, Sgeach. The hawthorn name occurs in many placenames in Ireland too. Mac Coitir (2003) lists 11 counties with placenames including 'Skeagh', and there are other variations on this placename including 'Skea', 'Skagh' and 'Ska'. He also mentions a most poetic placename, 'Skeheenarinky' ('*Sceichín an rinnce*'), meaning the "hawthorn of the dancing place", in Tipperary.

Throughout Ireland, there are many stories told of famous landmark trees, and a considerable number of them are hawthorns. One such was the famous Beggar's Bush area in the south city region of Dublin, which is said to have been named after an ancient hawthorn tree under which beggars and perhaps highwaymen would shelter before descending on Dublin city. The Beggar's Bush was situated on the corner of Shelbourne and Lansdowne roads in Ballsbridge, on land that became the Trinity College Botanic Garden, established there in 1806. Up to the mid-1900s, there was an old hawthorn tree growing in the Botanic Garden that was said by some to have been from the original Beggar's Bush (Wyse Jackson 1987).

While hawthorn is widespread as a native plant in many different habitats, it has been widely planted for several centuries to the extent that today we cannot be sure what would have been its native distribution. Hawthorn has been much used throughout Ireland as a hedging plant; it is appreciated as a stock-proof species. It is robust and thorny, and cattle and sheep are less likely to push through it when it forms a dense hedge than they are to push through other native plants that might be used surrounding a field or other enclosure. Hawthorn is therefore of considerable economic importance, particularly in its use for hedging, right up to the present day, as it provides an impenetrable barrier, especially if regularly pruned. As well as growing plants, cut branches of hawthorn, blackthorn and holly are regularly used to fill gaps in hedges to keep cattle and sheep in their fields.

From the earliest of times, Ireland was a country of smallholdings and small field-plots (Evans 1957). For example, before the famine, 80% of all Irish farms were less than 15 acres (6 ha) in size. Evans suggested that most hedgerows are of no great antiquity but were planted as a result of the agrarian revolution, in different parts of the country between 1750 and 1850, when thorn hedges were planted and stone walls were constructed on land where boundaries had been more temporary or ill-defined up until then.

Eileen McCracken (1971) wrote that "The present-day pattern of hawthorn hedgerows interspersed at regular intervals with a single tree was unknown. These hedges with their individual ash, elm, sycamore, chestnut, or oak were an innovation of the late seventeenth century and did not spread far beyond the Pale until the mid-eighteen century." The Pale was an area of land, centred on Dublin, that stretched from Dundalk in Co. Louth to Dalkey in south Co. Dublin which was the base and center of English rule in Ireland in medieval times. McCracken continued – "From the mid seventeenth century one finds increasing numbers of leases binding tenants not only to make a substantial hawthorn hedge around his holding but to plant and maintain trees in it."

Some field boundaries are of course very ancient – stone walls have been found buried deep within peat bogs, such as the extensive patterns of field boundaries, settlements and tombs dating from the Neolithic period at Céide

Fields in Ballycastle, north Co. Mayo (Boschiero et al. 2018). The complex hedge management practices found throughout much of Britain, including hedge laying, have been but a little part of the Irish landscape or standard traditional farming practices. Hedgerows were much used to provide firewood, particularly in poorer parts of Ireland, and would probably have been hacked and harvested in a largely uncontrolled manner. In some areas too, *Crataegus* would have been little planted for hedging, with field boundaries mainly consisting of a ditch and earth bank or mound, topped with wild brambles and gorse. Lucas (1958) described the way a gorse (or hawthorn) hedge was made in Ireland, with the seeds inserted in a twisted *súgán* rope of hay or straw, which was then laid on the top of a field margin embankment and covered lightly with soil.

Apart from its role in agricultural development, the presence of hawthorn in the Old Irish tree-list, demonstrates its importance and significance in early Ireland. *Crataegus monogyna* was one of the Class 2 trees, *Aithig Fhedo*: "commoners of the wood", recorded as '*scé*', along with alder, willow, birch, elm, wild cherry and rowan; less highly prized than Class 1, oak, hazel, ash, yew, holly, Scots pine and wild apple (*Airig Fedo*: the nobles of the wood), but one step up from the lower divisions of the wood, Class 3, that included elder, blackthorn, spindle-tree, whitebeam, aspen, juniper and arbutus. Class 4 trees, "the bushes of the wood" comprise bracken, gorse/furze, bramble, wild rose, broom, heather and bog-myrtle. Complete versions of the tree-list are found in *Bretha Comaithchesa*, the c.a. 8th century early law text on farming. The order of importance into four classes indicated the economic importance of these trees in early Ireland. Not a lot is said about the Class 2 trees. The *Bretha Comaithchesa* describes the penalties that apply to damaging one of the trees in this list that belongs to another person, where, for example, a milch cow or sheep would be due in compensation for damaging the tree, and penalties would increase in severity according to the amount of damage (Kelly 1997).

Hawthorn is also included in the Irish tree alphabet. There is a strong old Irish connection between the ancient Irish alphabet Ogham and Irish tree folklore. Ogham came into use in Ireland in about the 4th century and consisted of a series of markings that survive today cut into the edges of standing stones. Each letter consists of a series of markings and was assigned a name, often associated with an Irish tree. For example, B was *beith* (birch) and D was *dair* (oak). Originally, probably only eight letters were named after trees (birch, alder, willow, oak, hazel, ash, yew and perhaps pine). In the Middle Ages, other tree names were added to the remaining letters, resulting in a full tree alphabet (Coll 2000). Hawthorn represents the sixth consonant of the Ogham Alphabet, H, *hUath*, meaning "fear".

Hawthorn in folk medicine traditions

Hawthorn has long been recognized for its therapeutic value and medicinal qualities. Lady Wilde (1826–1896) was a great compiler of old Irish folklore and mythology and included information on the herbal value of hawthorn in her book *Ancient legend, mystic charms and superstitions of Ireland* (1991). She wrote that there are seven herbs of great value and power [in Ireland]; they are ground ivy, vervain, eyebright, foxglove, the bark of the elder-tree and the young shoots of the hawthorn.

> Nine balls of these mixed together may be taken, and afterwards a potion made of the bog-water and salt, boiled in a vessel, with a piece of money and an elf-stone. The elf-stone [an ancient stone arrowhead] is generally found near a rath; it has great virtues, but once lifted up by the spade it must never touch the earth, or all its virtue is gone.
>
> (Wilde 1991)

An earlier reference to the recognition of its medicinal value in Ireland is included in Caleb Threlkeld's (1726) book *Synopsis Stirpium Hibernicarum* in which he said "The Haws are accounted *Diuretick*, good for the *Stone*, *Gravel*, and *Pleurisy* [emphasis added]".

Their berries are very nutritious and are mildly sweet and have a tart, tangy taste. Hawthorn berries are used in traditional medicine in many countries and culture, such as Chinese traditional medicine. They have been widely used as an herbal remedy for hypertension (high blood pressure), cardiovascular and coronary artery disease and as a restorative after heart attacks (it is said to relax and dilate the coronary and peripheral arteries, which helps blood flow to the heart and improves blood circulation in general). It is also said to be a useful treatment for raised cholesterol levels and for digestive problems (indigestion, flatulence, colic and the sensation of feeling full). Chemical constituents of hawthorn include flavonoids, tannins, saponins, rutin and cyanogenic glycosides.

Hawthorn bark was steeped in black tea and the liquid that resulted was used to treat toothaches in Co. Leitrim and also as a remedy for burns (Allen and Hatfield 2004). Shaw (1998) noted its herbal use to treat heart ailments, including coronary artery disease, angina, hypertension and heart failure. Early European herbals recommend hawthorn for stomach colic and diarrhoea. It was used to treat diarrhoea in cattle too in Ireland – there is a record of its use to treat diarrhoea in bullocks in Co. Longford (Allen and Hatfield 2004). Devlin (2011) gives a warning on its use however, mentioning folklore from Tipperary that warns that eating haws will give you yellow jaundice.

The first mention of its effect on the heart is by Quercetanus, the personal physician of King Henry IV of France, who concocted an 'anti-age syrup' from the plant (Vogel/Bioforce AG 2012). O'Regan (1997) reported a story of a medical doctor in the late 19th century in Co. Clare who treated hundreds of patients with a secret remedy that turned out after his death to be a tincture made from hawthorn berries. He had used it to treat angina, coronary thrombosis, palpitations, chest pains and other heart disorders.

A tree of many uses

In addition to its use in traditional medicine, hawthorn has had a long history of use for many purposes in Ireland. Its timber is hard and was certainly used for firewood. For example, ancient charcoal samples of hawthorn wood have been identified at various sites associated with outdoor cooking in Ireland. I have read some lore that hawthorn wood was said to have been used by witches to make their broomsticks, and of course, they appear as magic wands in the Harry Potter books. In the books, hawthorn, "makes a strange, contradictory wand", referring to it as "full of paradoxes" between life and death, but this wood can help with "healing magic, but they are also adept at curses" (Rowling 2015). These are contradictions and paradoxes that are very much a part of Ireland's hawthorn folklore tradition too. A cut branch of hawthorn was used in some parts of Ireland as a simple harrow, dragged along the ground to remove the weeds in a field, or to help sow seeds (Lucas 1958). While it was and is still taboo to cut down a sacred tree, it is acceptable to cut the same species for example for firewood in a hedgerow. I have not found much related to the specific uses of its timber in Ireland but in Scotland, Milliken and Bridgewater (2004) mentioned that its wood has been used to make durable cogs and teeth of mill wheels and other machinery, and tool handles too, as it is tough and fine grained. The leaves of hawthorn have been used to dye fabrics a dark blue colour (Danaher 1964). The branches of hawthorn were burnt and made excellent ashes for bleaching cotton (Rutty 1772).

In ancient Ireland, hawthorn would have been used as a food source, even though its fruits are not particularly sweet or tasty. It crops heavily and is often more widely available later in the autumn than many other native fruits. Geraghty (1996) noted that haw seeds have been found in excavations of Viking Dublin dating back more than 1,000 years. The same author mentioned the old saying used in Ireland, "When all fruit fails then welcome haw," suggesting that it was probably a useful 'last resort' fruit. Haws are best collected late in the season when they become more palatable and a little bit sweeter. The seeds have also been found in medieval deposits in Kilkenny (Lyons 2016).

The fruits of the hawthorn are edible raw but are not particularly tasty and are more commonly made into jellies, jams and syrups and wine. Haw jelly is a delicious and delicately flavoured alternative to red currant jelly to serve with meat. To prepare jelly from the berries (haws), the fruits are trimmed of their stalks, washed, stewed and strained before being boiled with sugar and bottled. Haws can be mixed with other hedgerow fruits (rosehips, crabapples, blackberries and other fruits) to make 'hedgerow jam'. In Scotland, haws have been used to a limited extent for commercial purposes, for the preparation of jams and preserves, usually as part of fruit mixes (Milliken and Bridgewater 2004), but I have not heard of any such commercial use in Ireland. The fruits can be mixed with flour to make a nutritious bread, a good source of Vitamin C. For example, Hilton (2007) suggested that dried fruits can be ground and added to flour used in baking. The fruits are rich in antioxidants. Rutty (1772: 85) mentioned such usage and said that, "The ripe fruit is sold in our market [presumably in Dublin], and eaten by the poor, and reduced to meal, may be a substitute for bread in time of scarcity." Hawthorn flowers are also edible and can be added to fruit salad or used to flavour puddings such as custard and junket. Hawthorn buds and young leaves can be included in salads (Phillips 1983, recommends them in a salad with beetroot or potatoes). They have a nice, fresh flavour with very little bitterness if picked just as soon as they open from the buds in spring. Mears and Hillman (2007) suggested that the flesh of haws can be processed and used as food by crushing the fruits and removing the skins and seeds to leave an edible gelatinous mush. The flowers can be used to make wine and a liqueur (Ó Céirín and Ó Céirín 1978). Fruits have been used to make wine and can be added to brandy and other spirits to make liqueurs. Hawthorn blossom steeped in brandy with some added sugar makes an unusual liqueur. Young hawthorn leaves and buds were picked by Irish children and eaten, called 'Bread-and-cheese'.

Hawthorn – a rich folklore in Ireland

The folklore related to hawthorn in Ireland and elsewhere is rich and diverse. It is frequent both in popular folklore and in mythology. In Ireland, on one hand, the tree is revered for its beauty, utility and sacred connections; on the other hand it is approached cautiously because of its connections with death, the underworld, bad luck and mischievous or malevolent little people, the fairies. The image of a hawthorn tree standing alone in the center of a field surrounded by grass or a crop is a powerful symbol of the special nature of the hawthorn in Ireland, even today. One might expect that in these days of modern agriculture, that such impediments to the plough or combine harvester would be swept away. In addition to standing as lone trees in a field, fairy trees often acted as boundary trees for fields or corner posts, helping to define limits between one landowner and another. Of course, some may say these trees are retained as they can provide shelter for stock against the worst of Ireland's weather, but the reasons why such trees have survived go beyond any practical or rational reasons. Today those who have care for the land increasingly recognize that trees, wildlife and all forms of biodiversity are to be nurtured and appreciated, rather than cleared for any perceived small increase in agricultural yield. But even today in 'Modern Ireland' there is a recognition that there are cultural and superstitious reasons not to interfere with these trees. Many don't really believe the superstitions associated with the so-called Fairy Trees (or Fairy Thorns), but why take the chance?

I recently embarked on a totally non-scientific survey of the folklore about hawthorn that has been posted on the internet, delving into numerous sites and blog posts from Ireland or about Ireland to gather what I could about contemporary folklore associated with hawthorns as fairy trees. Nine pages of manuscript notes were the result. While the information provided was varied and inconsistent, and from time to time somewhat eccentric, some key themes emerged. First of all, lone hawthorn trees were generally associated with the fairies, known classically in Ireland as the '*Sídhe*', "wee folk" or "little people". These trees are said to be amongst their entry points to the 'otherworld' or meeting places. The many stories repeated that damaging or destroying such trees risks the wrath and revenge of the *Sídhe*, with a lifetime of bad luck, or worse (Dalton, 2018). Secondly, that they have a strong link to religious beliefs; the power of the hawthorn for healing still conveys widespread respect for the hawthorn. One blogger wrote: "My Dad RIP planted a whithorn (sic) tree and said 'there, its your tree since you're jinxed'. I was shocked and hurt by it. Not a luck person. Yes, I met the Ban sidh and she gave me a gift" (Sullivan 2021). Thirdly, the hawthorn should never be taken indoors, as it is a presage of death (Curtis and Whelan 2019).

Stories about the origins of the *Sídhe* in Irish mythology tell that when the Milesians or Gaels arrived in Ireland, they fought and eventually replaced the existing inhabitants, the Tuatha Dé Danann, who were defeated in the battles that took place. They became the *Sídhe* and retreated (or were banished) underground to the 'Other World'. Their entrances and gateways from the Other World were said to be these fairy trees, as well as burial mounds, underwater and fairy forts, all of which were protected by their powerful magic. It was said to be very perilous to sleep under a fairy tree or take a nap in a fairy fort because of the risk of being swept away to the Otherworld by the fairies.

Ring forts are, in reality, medieval or older farming homestead enclosures, perhaps where cattle or other stock were protected from cattle raiders or wolves (the last wolves in Ireland were recorded in the 1700s). The internet stories say that when people moved from these ring forts, the fairies made them their new homes. Hawthorn trees in or surrounding these forts are a very common sight today. As a child growing up in rural Co. Kerry, our home was overlooked by a ring fort on a promontory above us. The fort was dense with hawthorn, blackthorn (*Prunus spinosa*) and Ash (*Fraxinus excelsior*) trees. We always treated this fort with respect and understood that we were in a place of great antiquity and with a special spiritual and perhaps a somewhat menacing atmosphere.

Again and again, on the Internet, the same stories are told of those who suffered because of removing or interfering with a hawthorn fairy tree. A story about John DeLorean, the American car manufacturer is often repeated. In the 1970s, DeLorean set up a manufacturing plant for his company at Dunmurry, near Belfast in Northern Ireland. In 1982, workers from the plant claimed that the business suffered ongoing bad luck and was eventually closed because a 'faery thorn bush' had been disturbed during construction of the factory. It is even said that John DeLorean drove the bulldozer himself to remove the tree because none of the workers would touch the tree. Subsequently, the management were said to have taken the removal of the tree seriously and attempted to mitigate the bad luck by planting a similar bush.

Road building projects in Ireland are frequently said to have been rerouted because of fairy trees on the proposed path of a new road. One such is the celebrated Fairy Tree of Latoon near Ennis. This tree was said to be especially important because it was the meeting point of travelling 'wee folk' from all around the world, or in other versions of the story, where the Munster fairies met those from Connacht. The preservation of this tree was championed by Edmund (Eddie) Lenihan, an Irish author, storyteller and folklorist. In 1999, Lenihan stood up to road builders in Co. Clare who had wanted to cut down a special whitethorn tree and he succeeded in altering the road project to spare the tree. Shortly afterwards, unknown vandals tried to destroy the tree and cut off all of its limbs. However, it has since recovered and is growing well on the roadside in Latoon, Co. Clare.

Hawthorn, like so many folklore tales, has a lighter benevolent side and a darker side too. Stories have been told that young women who rinse their faces in the morning dew of hawthorn leaves on the 1st of May will always have a beautiful complexion "and for ever after handsome be". Hawthorn flowers were given to newly married couples for good luck. Nevertheless, other folklore warns against bringing hawthorn blossom into the house, because it is unlucky and would cause a death in the family.

Bealtaine is the traditional Irish festival recognizing spring at its peak and welcoming the arrival of summer. It was (and to some extent still is) celebrated in Ireland, Scotland and the Isle of Man. The Bealtaine Festival is regarded as a classic Celtic festival, halfway between the Spring Equinox and the Summer Solstice, and exactly six months away from Samhain on 1st November, the Celtic New Year. My informal Internet survey told me that on 1st May, traditionally, doors and windows outside homes were hung with hawthorn branches to ward off evil and bad luck. Spring was a time when new milk from cattle was welcomed following the birth of calves, and it was feared that the milk and butter would be stolen by the fairies. The hawthorn decoration was thought to act as a protective talisman against the fairies.

Biddy White Lennon and Evan Doyle (2013) write about hawthorn in their book on wild food: "Since Celtic time [it] has been the flower of The Spring Festival, marking the start of the farming year – La Bealtaine. In Irish folklore May Eve and May Day is a time when the sidhe, the fairies, are about, as excited as humans at the coming of new milk and butter and liable to steal farm produce for themselves – one reason why people left lone hawthorns for the fairies."

The eve of Bealtaine is a night associated with malevolent activities by the fairies and amongst some rural people, is still regarded as a night to stay at home safely indoors. At the festival of Bealtaine, 'it is said that' women would hold a twig of hawthorn blossom to attract a husband. Sprigs of hawthorn blossoms in the hair at weddings, or in bouquets, were to symbolize a union of love. Mac Coitir (2003) described how women hoping to marry would hammer nails into a hawthorn tree growing in a stone circle at Longstone Rath, near Naas in Co. Kildare. Wilde (1887) and Wood-Martin (1902) described an account of a marriage in Co. Kerry, in 1830, written about by Lady Wilde:

> A large Hawthorn tree that stood in the middle of a field, near a stream, was hung all over with bits of coloured stuff, while lighted rush candles were placed here and there amongst the branches, to symbolize, no doubt, the

> new life of brightness preparing for the bridal pair. Then came a procession of boys marching slowly with flutes and pipes made of hollow reeds, and one struck a tin can with a stick at intervals, with a strong rhythmical cadence. . . . A boy followed, bearing a lighted torch of bog-wood. After him came the betrothed pair hand-in-hand, a large square canopy of black stuff being held over their heads. . . . The procession then moved on to a bonfire, evidently the ancient altar, and having gone round it three times, the black shroud was lifted from the bridal pair, and they kissed each other before all the people, who shouted and waved their branches in approval.
> (Wilde 1887: 147)

Outside the home, a hawthorn bush was transformed into a May Bush at Bealtaine, decorated with seashells, eggshells, trinkets and other materials. The strips of cloth and decorations were to appease the fairy guardians who live in it. May poles, an important part of spring folklore traditions in England around which people would dance, were originally made from hawthorn.

Beyond their use for a May Bush, many lone hawthorn trees have been transformed into 'Rag Trees' (or 'Wishing Trees') throughout Ireland. These will often occur beside a holy well, or at an ancient site of some antiquity, often with religious connections. The rags originally were usually cloth strips taken from the clothing of a sick person and tied to the tree as a petition to a local saint or deity. As the cloth rotted away, so did the sickness. People left a wide variety of items – prayers written on small folded pieces of paper, personal tokens and gifts (such as toys, socks, beads, photos, balloons, ribbons, strips of fabric and so on), in the hope of being healed, attaining good fortune or having a prayer answered. These decorated trees continue today and are relatively common throughout Ireland. Today, the materials tied on to the trees are often made of plastic and not biodegradable, so we cannot be sure of the impact this may have on the efficacy of their prayers! Lucas (1963) surveyed 210 holy wells in Ireland and recorded their associated trees. One hundred and three had hawthorns growing beside them, 75 had ash trees and the rest were random species.

While the May tree is associated most closely with the fairies, today these rag trees are more connected with Christian religious sites and the prayers that visitors and pilgrims leave for themselves or loved ones. Pagan and Christian traditions have become intermingled and the hawthorns (and blackthorns) in church grounds are both symbolic of pagan and Christian traditions, representing an interesting crossover between two systems of belief and practice. One can see this also at holy wells, baptismal sites and water springs too, where these trees are (or were) revered for their spiritual qualities and physical healing properties. There are an estimated 3,000 holy wells in Ireland (Lysaght 1979).

These important shrubs have also been called 'Mass bushes', and although they are generally hawthorns, other species can be found, sometimes ash and more rarely oak, willow, elder, holly, rowan, alder, elm, yew and [non-native] fir (Synnott 1979). A link for the Christian tradition was a story that Christ's crown-of-thorns worn at the crucifixion was made from hawthorn. The cross on which Christ was crucified was also said to have been made of hawthorn wood (Mac Coitir 2003). This connection between the hawthorn and Christianity helped to strengthen its link with the Christianity, emerging from paganism and its important healing power.

A.C. Haddon and C.R. Browne (1894) refer to an interesting tradition from the Aran Islands, Co. Galway. They wrote: "Pin-wells and rag-bushes are still frequented, and on the night before emigrating people will sleep in the open, beside one of the holy wells, in order that they may have good fortune in the country to which they are going."

This link between the hawthorn and Christian stories is well illustrated too by the 'Glastonbury thorn' that is associated with a supposed visit of Joseph of Arimathea to England in the 1st century. Joseph of Arimathea was responsible for the burial of Jesus after his crucifixion. The legend is that when he stuck his staff in the ground at Glastonbury in Somerset, England, it took root and grew into a tree. This tree, a hawthorn, is said to bloom only on Christmas Day, and in other stories, to bloom twice in one year. There are several trees growing adjacent to Irish holy wells that are said to have grown from walking sticks belonging to a variety of saints. One example is in Listerling parish in Co. Kilkenny, where a thorn tree is said to have grown from St. Moling's walking stick (Mac Coitir 2003).

Another good example of the intermingling of stories in Irish folklore is about the Irish Leprechaun that buries his crock of gold at the end of a rainbow. The Internet tells me that "a popular story among Irish children is that the 'wee folk' (fairies) would hide their pots of gold near the trunks of hawthorn trees" (Chavis and MacDonald 2022). In order to see the gold, it was necessary to go to the tree at the stroke of midnight and sit on a three-legged stool (made of ash wood).

Hawthorn blossoms have an associated folklore that is worth exploring too. The Internet tells me that these blossoms "are said to be highly erotic to men". Another writer said that the creamy white blossoms have a hypnotic, faintly repugnant scent, said to smell like the Great Plague of London – the bubonic plague of 1665 to 1666. Nevertheless, the blossom "is said to aid love and fertility, as long as it is kept outside!" (Schneidau 2019). The smell of the flowers, to attract insects for pollination, has a faintly 'carrion-like' aroma, caused by the chemicals they release, including triethylamine. Triethylamine is one of the first chemicals produced by a dead human body when it begins to decay, and gangrene has a similar odour. The aroma of hawthorn blossom is also associated with sex and said to possess the musky smell of semen (McGarry 2020).

Finally, a well-known term in Irish history is the 'hedge school', created particularly during the period when Britain's penal laws were in effect, from 1723 until 1782, directed at Roman Catholics. At that time, these laws prevented Roman Catholic schools from operating. While the term 'hedge school' suggests that these were convened outdoors, in reality, the basic education they provided was generally indoors, in a house or barn, so sheltering under the shade of a spreading hawthorn is more mythology than fact. The hedge schools taught basic writing, reading, grammar and maths, in Irish and English, and occasionally Latin too. During the 1820s, they received up to 400,000 students each year but declined in the 1830s with the formation of the National School system (Adams 1999).

Conclusions

One of the richest sources of manuscript information on folklore in Ireland during the last century is the remarkable Schools' Folklore Scheme, now known as the Schools' Manuscript Collection, which ran from July 1937 until December 1938. This scheme was initiated by the Irish Folklore Commission, in collaboration with the Department of Education and the Irish National Teachers' Organisation by National Schools in the Republic of Ireland. Some 100,000 children in 5,000 primary schools throughout the 26 counties acted as folklore collectors, gathering information and stories from their parents, grandparents and neighbours. The results of the scheme included about half a million pages of manuscript, which are now bound into 1,126 volumes housed in the library of the National Folklore Collection at University College Dublin (UCD). When I searched the Schools' Manuscript Collection for the words 'hawthorn' and 'whitethorn', I found 746 entries for the former and 701 for the latter, from almost every corner of Ireland. A further 149 entries were found when I searched its Irish name '*sceach*'.

At a time when traditional knowledge on plants and their uses is disappearing fast worldwide, it is heartening to note that there is still a rich lore of stories about a plant such as hawthorn living on in contemporary Ireland. Much more is there for research amongst the wealth of historic manuscripts and literature in Irish and English, helping to ensure that future generations will not lose the lore, knowledge and stories on which our present culture has been built. For all the reasons given in this article, it is clear that hawthorn is one of the great treasures of the Irish landscape, a genetic resource of inestimable value, past, present and future, and a fascinating topic for future research. In the Irish Schools' Manuscripts alone, I have a further 1,596 stories still to read!

References

Adams, J. R. R. 1999. The Hedge Schools and Popular Education in Ireland. In *Irish Popular Culture 1650–1850*, edited by J. Donnelly and K. Miller. Irish Academic Press, Dublin, Ireland.

Allen, D. E., and G. Hatfield. 2004. *Medicinal Plants in Folk Tradition: An Ethnobotany of Britain and Ireland*. Timber Press, Portland and Cambridge.

Anonymous. 2005. *Champion Trees: A Selection of Ireland's Great Trees*. Tree Council of Ireland, Dublin.

Boschiero, Patrizia, Luigi Latini, and Seamas Caulfield, eds. 2018. *The Céide Fields, Ireland: International Carlo Scarpa Prize for Gardens 2018*, 29th ed. Fondazione Benetton Studi Ricerche, Treviso, Italy.

Chavis, Lisa, and MacDonald, Cheryl. 2022. *Fairy Trees in Ireland. In What Boundaries? Live your Dream!* Available at https://whatboundariestravel.com/fairy-trees-ireland/

Colgan, Nathaniel. 1904. *Flora of the County Dublin*. Hodges, Figgis & Co. Ltd, Dublin.

Coll, Marian, ed. 2000. *Our Trees: A Guide to Growing Ireland's Native Trees in Celebration of a New Millennium*. Coillte and Woodlands of Ireland, Dublin.

Curtis, Tom, and Paul Whelan. 2019. *The Wild Food Plants of Ireland*. Orla Kelly Publishing, Ireland.

Dalton, Angela. 2018. *Hawthorn-Tree of the Wee Folk*. Garden Grim. https://gardengrimoire.wordpress.com/2018/03/06/hawthorn-tree-of-the-wee-folk/

Danaher, Kevin. 1964. *Irish Customs and Beliefs*. Mercier Press, Cork (reprinted 2004).

Devlin, Zoë. 2011. *Wildflowers of Ireland: A Personal Record*. The Collins Press, Cork.

Evans, Estyn E. 1957. *Irish Folk Ways*. Routledge & Kegan Paul Ltd, London.

Geraghty, Siobhán. 1996. *Viking Dublin: Botanical Evidence from Fishamble Street*. National Museum of Ireland, Medieval Dublin Excavations 1962–81, Ser. C, Vol 2. Royal Irish Academy, Dublin.

Grigson, Geoffrey. 1975. *The Englishman's Flora*. Paladin, Frogmore, St Albans, Herts, UK.

Hackney, Paul, ed. 1992. *Stewart & Corry's Flora of the North-East of Ireland*, 3rd ed. Institute of Irish Studies, The Queen's University of Belfast, Belfast.

Haddon, A. C., and C. R. Browne. Anon. 1894. Review of paper by A. C. Haddon and C. R. Browne on The Ethnography of the Aran Islands, County Galway. *Nature* 49(1272), March 15, 1894.

Hilton, Jonathan. 2007. *Wild Food for Free*. Hamlyn, London.

Kelly, Fergus. 1997. *Early Irish Farming*. School of Celtic Studies, Dublin Institute for Advanced Studies, Dublin, Ireland.

Lucas, A. T. 1958. Furze: A Survey and History of Its Uses in Ireland. *Béaloideas* 26: 1–204.

Lucas, A. T. 1963. The Sacred Trees of Ireland. *Journal of the Cork Historical and Archaeological Society* 68: 16–54.

Lyons, S. 2016. Food Plants, Fruits and Foreign Foodstuffs: The Archaeological Evidence from Medieval Ireland. In *Food and Drink in Ireland*, edited by E. FitzPatrick and J. Kelly, pp. 111–166. Royal Irish Academy, Dublin.

Lysaght, Patricia. 1979. Trees in Irish Folk Tradition. *Irish Biological Records Centre Newsletter* 14(May 1979): 22–39.

Mac Coitir, Niall. 2003. *Irish Trees, Myths, Legends and Folklore*. The Collins Press, Cork.

McCracken, Eileen. 1971. *The Irish Woods since Tudor Times: Their Distribution and Exploitation*. David & Charles, Newton Abbot, UK.

McGarry, Marion. 2020. *Galway Mayo Institute of Technology – Death, Sex, Superstition and Fear: The Hawthorn Tree in Ireland*. Available at RTE.IE/brainstorm/2020/0506/1136776-hawthorn-tree-ireland-folklore/.

Mears, Ray, and Gordon Hillman. 2007. *Wild Food*. BBC and Hodder & Stoughton, London, UK.

Milliken, William, and Sam Bridgewater. 2004. *Flora Celtica: Plants and People in Scotland*. Birlinn, Edinburgh.

Ó Céirín, Cyril, and Kit Ó Céirín. 1978. *Wild and Free: Cooking from Nature*. O'Brien Press, Dublin.

O'Regan, Paula. 1997. *Healing Herbs in Ireland*. Primrose Press, Blackrock, Co., Dublin.

Phillips, Roger. 1983. *Wild Food*. Pan Macmillan, London.

Rowling, J. K. 2015. *Wand Woods*. Pottermore. Available at https://www.wizardingworld.com/writing-by-jk-rowling/wand-woods

Rutty, John. 1772. *An Essay Towards a Natural History of the County of Dublin*, vol. 1. W. Sleater, Dublin.

Schneidau, Lisa. 2019. *The Fairy Trees: Blackthorn, Hawthorn, and Rowan*. Folklore Thursday. Available at https://folklorethursday.com/folklife/the-fairy-trees-blackthorn-hawthorn-and-rowan/

Shaw, N. 1998. *Herbalism: An Illustrated Guide*. Element Books, Shaftesbury, Dorset.

Sullivan, Sheila. 2021. Comment on *The Warrior's Battle Frenzy*. Ali Isaac Storyteller. Available at https://www.aliisaacstoryteller.com/post/riastradh-the-warrior-s-battle-frenzy

Synnott, Donal M. 1979. Folk-lore, Legend and Irish Plants. In *Irish Gardening and Horticulture*, edited by C. Nelson and A. Brady, pp. 37–43. Royal Horticultural Society of Ireland, Dublin.

Threlkeld, Caleb. 1726. *Synopsis Stirpium Hibernicarum*. F. Davys et al., Dublin.

Vogel, A./Bioforce AG. 2012. Fresh from the Nature: A. Vogel Plant Encyclopaedia. *Crataegus monogyna* Jacq. & *laevigata* (Poir.) DC Hawthorn. Available at www.avogel.ch/en/plant-encyclopaedia/crataegus.php. Accessed on August 23, 2012.

White Lennon, Biddy, and Evan Doyle. 2013. *Wild Food: Nature's Harvest: How to Gather, Cook & Preserve*. The O'Brien Press, Dublin.

Wilde, Lady. 1887. *Ancient Legends, Mystic Charms and Superstitions of Ireland: With Sketches of the Irish Past*. Ticknor and Co., London

Wilde, Lady. 1991. *Irish Cures, Mystic Charms & Superstitions*. Sterling Publishing Co. Inc., New York.

Wood-Martin, W. G. 1902. *Traces of the Elder Faiths of Ireland: A Folklore Sketch. A Handbook of Irish Pre-Christian Traditions*, vols. 1, 2. Longmans, Green, and Co., London.

Wyse Jackson, Peter S. 1987. The Botanic Garden of Trinity College Dublin 1687 to 1987. *Botanical Journal of the Linnean Society* 95: 301–311.

Wyse Jackson, Peter S. 2014. *Ireland's Generous Nature: The Past and Present Uses of Wild Plants in Ireland*. Missouri Botanical Garden Press, St Louis, MO.

9 The carob as a blessed and accursed tree

Amots Dafni and Saleh 'Aqil Khatib

Introduction

Trees have always been regarded as the first temples of the gods, and sacred groves as their first place of worship, and both were held in utmost reverence in the past (Pliny 1945: 12.2.3; Quantz 1898: 471; Porteous 1928: 190). Thus, it is not surprising that individual as well as groups of sacred trees have been a characteristic of almost every culture and religion that has existed in places where trees can grow (Philpot 1897: 4; Quantz 1898: 467; Chandran and Hughes 1997: 414). It is not uncommon to find traces of tree worship in the Middle East, as well. However, as William Robertson-Smith (1889: 187) noted, "there is no reason to think that any of the great Semitic cults developed out of tree worship". It has already been recognized that trees are not worshipped for themselves but for what is revealed through them, for what they imply and signify (Eliade 1958: 268; Zahan 1979: 28), and, especially, for various powers attributed to them (Millar et al. 1999: 35; Hamilton 2002: 57).

Pamela R. Frese and S. J. M. Gray note:

> Sacred trees have a ritual significance. The trees and their meanings may be incorporated into rituals of curing, initiation, marriage, and death. Trees used in any of these contexts stand for the divine and represent the sacred beliefs being honored through the ritual.
>
> (1995: 32)

In the same vein, Canaan adds:

> The present day peasant does not venerate the trees themselves but the divine power which acts in them and which is derived from a godly person whose soul is supposed to be inhabiting the shrine, tomb, cave or spring with which they became associated. Often these holy men have appeared in the tree itself or near by.
>
> (1927: 151)

DOI: 10.4324/9780429320897-11

Regarding this point, it is pertinent to cite John Mills who declared: "Yet in no country are the people more awed by trees than in Palestine" (1864: 54).

The very attitude toward a certain plant as blessed (of "esteemed") or accursed, can be subject to various interpretations during time and across religions. According to Christian traditions, the trees that assisted the Holy Family in their flight from the forces of Herod to Egypt are considered as blessed, while trees that were negative to Jesus or "assisted" his enemies are considered as accursed (Skinner 1913: 132, 145, 232). In Jewish sources, blessings are made over certain plants, such as the four species on Sukkot, or on aromatic plants during the Havdalah ceremony, but there are no plants that are explicitly blessed nor are there any plants that are "accursed".

According to what is written in the Quran, God swore by the olive and the fig trees (Surah 95, verses 1–4): "I shall swear by the figs and olives, and by Mount Sinai, and this safe stretch of land! We have created man in the image of great beauty". And as it says in Surah 24, verse 35: "And the lamp is made of glass, and the glass is like a shining star: it is illuminated by (the oil of) a blessed tree, the olive tree". In his ascent to heaven (the Night Journey) on the 27th of the seventh Muslim month called Rajab, the Prophet Muhammad arrived at a lote tree:

> And he certainly saw him in another descent at the Lote tree of the utmost boundary. Near it is the garden of refuge where he was concealed by the Lote tree which covered it. He did not swerve his sight from it or let it wander.
>
> (Surah 53, verses13–17, quoted from: سدرة المنتهى [the lote tree of the utmost boundary])

This refers to the area beyond which no one can reach, neither angels nor men. Some interpret it as referring to a lote tree standing in the highest part of the Garden of Eden, in the seventh heaven, on the right side of the throne of God. Therefore, these three species of tree are considered as more blessed than all the other trees.

Muslims in Israel refer to certain tress as "sacred trees", which in most cases have been sanctified as a result of miracles or wondrous events related to the lives of the "Wellis", the righteous ones whose souls dwell in the tree (Dafni 2006, 2011). Druze religious leaders often name venerated trees as "blessed trees" and consider that holiness can be attributed only to those individuals from whom the tree will accept their blessings (Dafni 2010: 51–52). In this article, we will deal with the status of the carob as a "blessed" or an "accursed" tree in the various traditions associated with the Holy Land.

Methods

Our fieldwork was conducted between 1999 and 2006 and centered on 35 Muslim and Druze villages in the Galilee. The survey covered 158 informants, comprising 47 Druze and 111 Muslims. Out of the all the informants included in this survey, only 29 interviewees, all of whom are mentioned in the text, expressed special comments on the carob (all the informants indicate the importance of the carob as an important food). They were 27 males and two females – in general women are reluctant to be interviewed, and when they agreed, the interview was held in the presence of other family members. The average age of the informants was 57.7 (+/-14.8) years.

Data collection

If the informant was not Hebrew speaking, the interview was held in Arabic with the help of an experienced interpreter. At the end of each interview, we read our note in the informant's presence to verify the validity of the data gathered. Specific questions included: Are there any stories and proverbs related to the carob; Does carob has any supernatural powers; Is it a "good" or "bad" tree; What is the informant's personal view as well as spiritual attitude toward this tree?

Results and discussion

The carob as a "blessed" tree

Druze religious leaders often name venerated trees as "blessed trees" and consider that holiness can be attributed only to those individuals from whom the tree will accept their blessings (Dafni 2010: 51–52). This was well expressed by our participant, the Druze sheikh, Kassam Bader, who oversaw the sacred site of Nabi Sabalan in Hurfeish: "Holiness is always accompanied by a blessing, and the blessing comes from the Prophet". Sheikh Hussein Shahin, a prominent religious leader from Beit Jann, adds that:

> The tree is blessed, and the man is holy, the tree belongs to mankind and the holy ones are only God and the prophets. The blessed trees are a monument to particular personages in the history of the Druze religion. According to our religion, no marks or offerings are made on their graves because for these special people, the blessed trees are the sole commemoration of their endeavors.
>
> (12.9.00)

Among the Druze, the carob tree enjoys much prestige, and therefore, one can find blessed carob trees in several sacred places such as the Carob Dome in Julis, al-Mubaraka in Kasra, and Makam el-Sheikh al-Fadel in Samia (all of which are in Upper Galilee). The most widely cited knowledge about the tree is captured in the following statements: "A good tree, people used to sleep under this tree" (Mustafa Halawi, 15.12.03, Usfiya; Tewfiq Amasha, 12.12.01, Mas'ada); "The carob is a good tree, important people used to sleep under the carob tree" (Ziad Talal, 3.10.03, Sajur); "The whole village sleeps under the carob tree and nothing happens to them" (Sa'id Muhammad, 25.5.03, Yanuh; Salman Abu Rukan, 12.1.03, Usfiya); "The carob tree is always blessed" (Kassam Bader, 9.12.05, Nabi Salaban); "Every fruit tree is blessed" (Hussein Shahin, Beit Jann, 12.9.00).

In the Druze village of Kisra, there is a blessed carob tree that bears the name of Sitt Sarah, who was one of the religious propagators of the Druze faith. Until the middle of the 20th century, the elders of village used to gather under the tree every Sunday and Thursday to discuss religious matters. Near the village of Samia, there was an ancient carob tree and an ancient oak tree that had dried up a few years earlier. The place is called "Al-Mubaraka" (the blessed).

Here and there one can find sacred carob trees, even in the Muslim sector. Near the village of Fureidis at the foot of the Carmel range, there was a carob tree dedicated to Sheikh Saleh. During the years of drought, the villagers used to pray for rain near the tree, and when the prayers were over, "rain would fall for a few days" (Abu Mahmud Anbus, 12.3.07, Fureidis).

At the entrance to I'bllin there was a carob tree dedicated to Mar Elias which was called the "Carob of Mar Elias" (Elijah the Prophet). The sister of the infor mant saw the Prophet Elijah near the tree, which he used to visit it occasionally. Whoever slept under the tree did not have any desire to work. The scent of the tree induced sleep and old people warned that one should not sleep under this carob tree because it causes idleness (Abu Fahad Raji Ajami, 16.8.04, I'bllin).

The sacred carob trees in Peki'in are called the "carobs of the sons of Jacob" and wonders and miracles are attributed to them. The Jewish residents of the village associate the carob trees in Peki'in with Rabbi Shimon Bar Yohai and the Christians associate them with the sons of Jacob. The miraculous powers of these trees are related in a legend:

> .Sons of Ras Abed used to conscript people from among the residents of Peki'in, and every year they would come and seize one of the young girls of the village. Once, when it was the turn of a certain girl, a delegation of tax collectors arrived at her house which was near some carob trees. The young girl escaped through the window and hid among the nearby carob trees while the tax collectors chased after her. When the young girl saw that the pursuers were coming closer, she raised her voice in prayer and called upon God to save her from the villains and begged the carob trees for protection. Her prayers were answered, and tongues of fire burst out from the carob trees of the sons of Jacob and burnt all those who tried to approach the girl. The flames followed the other members of the delegation and reached their clothes until they fled for their lives far from the lands of Peki'in. So, from that day onwards, conscription was annulled thanks to the carob trees of the sons of Jacob.
>
> (Hadad 1981: 132)

James George Frazer, sums up this subject by saying that most of the large trees (referring specifically to the oak trees in the Galilee) are inhabited by spirits called the daughters of Jacob. Apparently, this is a relic of idol worship that the Muslims are trying to uproot, but it is not possible to root them out of the hearts of the masses (Frazer 1923: 323). In Asia Minor and in Syria (during the time of writing, the Land of Israel was regarded as part of Syria) the carob tree was dedicated by Muslims and Christians to St. George, and sometimes his sacred shrines can be found under carob trees (Hehn 1885: 312). In Cyprus, there is also a church named after St. George of the Carob Trees (Howell 1970: 179).

In praise of the carob tree

Jewish sources treat the carob tree with great honor. Rabbi Zvi Schwartz excelled in doing so when he wrote an article entitled "The Carob is the Jewish Tree" (1995) in which he lists its praises. The carob tree symbolizes modesty and being satisfied with little:

> Every day a heavenly voice calls out and says: The whole world is nourished for the sake of Hanina my son, and my son Hanina is satisfied with a kav [ca. 1–1.79 liter according to various sources] of carobs from one Sabbath eve to the other.
>
> (Babylonian Talmud, Ta'anit 24:2)

Rabbi Yohanan talks about the juiciness and sweetness of the carob fruit in the following story: "I remember that when a child broke a carob, a thread of honey ran along both his arms" (Babylonian Talmud, Baba Bathra 71:1).

Rabbi Hanina relates: "When I came here (to the Land of Israel) I took my belt and the belts of my son and my donkey to circle the trunk of the carob tree in the Land of Israel, but they were insufficient" (they still could not go all around the trunk). He cut into one carob and drew out a handful of honey (which dripped from the carob) (Babylonian Talmud, Baba Bathra 91:2).

Rabbi Yehuda Leib Halevi Ashlag (1884–1954), in his commentary on the Book of the Zohar, comments on the importance of the carob as a fruit-bearing tree:

> As Rabbi Yossi says, trees from which much wisdom appears, that is to say the carob tree, the palm tree, and the pistachio which is a kind of nut . . . all the trees that bear fruit . . . each tree has a special path which it illuminates.
>
> (Shmot p. 5, letter 274)

In the opinion of botanists, the carob is not regarded as long-lived, and one rarely finds a carob tree that is more than a hundred years old (Prof. Simcha Lev-Yadun, 6.8.20, personal information). Some claim that the carob tree may reach an age of up to 300–400 years (Ghrabi 2005: 69), but they do not present any data to support these figures. According to Rabbi David Altschuler of Prague (1687–1769), one of the important commentators of the Prophets and the Hagiographa, the carob is a symbol of long life: "For the days of the tree which endures for a long time, such as the carob tree, and may the days of my people be like the days of this tree" (commentary on Isaiah 65, Altschuler 1776).

Is the carob tree accursed?

In Arab villages, carob and fig trees are regarded as the abodes of evil spirits and demons, as in the proverb which says: "The carob and fig trees are the dwelling place of Satan" (خروب وتين مسكن الشياطين). Some people believe that the reason for this is due to their black fruits. In Israel, it is recommended for the sake of safety not to sleep under a carob tree. An Arabic proverb mentions this explicitly: "Sleeping under a carob tree is not desirable" (النّومة تحت الخرّوبة مش مرغوبة) (Dafni and Khatib 2017: 52–53).

Tawfiq Cana'an explains this by saying that carob and fig trees have black fruit and are the preferred dwelling place for demons who gather under them for their meetings and their nightly processions. According to the same source, dreams about the carob tree mean catastrophe, destruction, death, and loss since the Arabic word "*haraba*" means "to destroy". Therefore, the farmer will not tie his donkey to a carob tree before asking permission from the jinn (Canaan 1927: 37).

In Arabic folklore, the carob tree is regarded as a cursed tree, the place where evil spirits (*jinns*) reside, and which brings bad luck (Canaan 1927: 37). According to testimony from the village of Artas near the Pools of Solomon, it is extremely dangerous to sit under a carob tree, especially at night, because it is contaminated by evil spirits. Once, during a funeral, a strange European brought some carob leaves. The villagers were struck with fear and asked how he could lay leaves on the grave that had blood on them (the veins of the leaves are red) (Crowfoot and Baldensperger 1932: 113). A similar tradition can also be found in Italy where they believe that the carob tree attracts lightning, and that the tree is inhabited by demons. In addition, they believe that whoever falls from a carob tree will surely die. For that reason, whoever rests under a carob tree is regarded as a bad person (Malossini 2013: 187).

When we asked our informants why it was "forbidden" to sleep under a carob tree, we received, among others [all are Muslims], the following answers: "The carob tree is the tree of demons and it is preferable not to sleep or to urinate under it because demons will leap upon you. The tree is accursed" (Sheikh Nur Rifa'i, 16.6.00, Majdal Krum); "Demons gather under the carob tree, and God punishes them with lightning, therefore large carob trees attract lightning" (Ibrahim Kadah and Adel Abu Hamidat, 21.12.03, Kfar Manda; Sheikh Abu Muhammad Hussein Almaz, 1.11.11, Ka'biyya); "A black angel dwells under the carob tree and therefore it is not good to sleep under it. It is forbidden to place a carob tree near the house because it contains demons" (Suleiman Shibli, 1.11.04, Shibli; Fatma Nufel, 21.11.11, al-Hilf); "Sleeping under a carob tree is prohibited because it has a deep shadow and the jinn lives inside the carob tree" (Jmil Abu Ra'id, 23.9.04, Mash'had); "Many people have slept under the carob tree and as a result became insane" ('Atef Mansour, 13.5.03, Kaukab Abu al-Hija; Nasser Halil, 1.1.05, Sakhnin); "There is a deep shadow under the carob tree, and the jinn (demon) is within it" (Abu Amin Haldi, 26.8.04, Khawled); "The carob is a tree of demons" ('Ataf Suleiman, 19.10.14, Eilabun); "The carob tree is full of demons" (Mahmud Zir'eini, 22.11.04, Tur'an); "Whoever sleeps under a carob tree will fall into a deep slumber. The scent of the carob induces sleep and then there is no desire to work, and this causes idleness" (Hamed Abu Mustafa, 31.12.03, Arrabe).

The carob tree and snakes

In Arab folklore, the carob tree is regarded as an accursed tree and the favorite dwelling place of snakes. When we asked why it was forbidden to sleep under the carob tree, we received the following replies: "Snakes love carob trees – they love the shade and the moisture in the heart of the tree" (Kobasi Sawad, 17.7.13, Hussniyya, near Ma'ale Tzviya); "Do not sleep under a carob tree, a snake may be hiding there; birds love the tree because of the shade and its size, and therefore the snakes go there. The trunk is dark colored and provides good cover for snakes" (Umm Mahmud, 2.8.04, Ma'ale Tzviya); "There is always the suspicion of snakes in the heart of the carob tree, and therefore it is not the custom to sleep under it; bees also nest in the carob tree" (Suleiman Salah Hutbe, 21.12.03, Arraba); "The carob tree attracts birds because of the sweet fruit and then the snakes come" (Hanna Shahadeh, 19.10.14, Mghar); "Snakes love carobs, and the snake makes his home in the carob tree or near it. Wherever there are carob trees there are snakes as well" (Bassam Hajj, 26.3.06, Rameh).

This means that whoever desires to sleep under the carob tree (no one doubts that it has a pleasant shade) risks an unexpected encounter with demons, and if these did not attack him badly, the snake will do so instead, and if he survives all the anticipated calamities, there is still the concern that he will be struck by lightning. The obvious conclusion is, of course, to search for better and safer places to sleep. Tawfiq Cana'an points out this glaring contradiction between the carob tree as having an excellent fruit (especially in bad times), and giving shade and refreshment throughout the year, and between being a accursed tree (Canaan 1927: 36). He says that demons dwell only in certain trees, and mainly in the carob tree. But the fact that there are sacred and blessed carob trees indicates that prophets and righteous men have the power to overcome evil forces for the benefit of faithful believers. In his words:

> I have never heard that pieces of cloth were hung on a tree [this was the practice for sacred trees, mostly for the fulfillment of personal requests] that was considered as the abode of demons. Besides, every kind of tree can be sanctified by a *wali* (a righteous man).
>
> (Canaan 1927: 37)

The demon in the carob tree

A medieval Jewish folk tale entitled "The Demon in the Carob Tree" tells the following story: Once there was a Hassid who was secretly God-fearing. He had one field in which there was a single carob tree with idols for a heathen worship under it. All the passersby used to sit under that carob tree which caused the seeds planted there by the Hassid to be lost. The Hassid said to his wife: "I will go and cut down that tree so that we do not lose our seeds". His wife told him: "Do as you wish". The Hassid immediately took up his axe and went out to chop down the tree. As he started cutting it down, a demon came out of the tree and said to the Hassid: 'Do not cut down the tree, and I will give you one golden dinar every day". The Hassid replied: "No!" (The demon then said) "I will give you three golden dinars". The Hassid left the tree alone and went home. Next morning, he returned to the tree and found three golden dinars. He found this every day until he became rich, built houses, and bought male and female slaves. He did not know from where the three gold coins were coming. In the course of time his sons and his slaves began to die. When he saw this, he said: "I am a sinner, because my sons and slaves have died". What did he do? He went with his wife to the tree to pick up the gold coins as he always did every day, and noticed people playing under the tree. He and his wife then began playing with them, but they did not find any gold coins. When the Hassid tried to cut down the tree, a goat came out from under it and told him: "'If you had not played and enjoyed yourself under the tree, I had three gold coins to give you as every day. Now that you have enjoyed yourself with them, I do not need to give you your reward". The Hassid attempted again to cut down the tree, but the goat told him: "Do not cut it down, else I shall kill you". The Hassid took fright and went to the Sanhedrin to tell them the story. They instructed him: "Go and sell all that you have bought with the dinars that the demon gave you, return them to him, and then cut down the tree". The Hassid went and did as he was told. He called his slaves and ordered them to cut down the tree. The goat then came out to face him and said: "Take six gold coins from me every day and do not cut it down". The Hassid replied: "If you give me all the silver and gold in the world, I will not assent to you". The goat fled away instantly, and the Hassid cut down the tree and went home. The next year, he planted the field and had a hundred-fold harvest which he sold for eight hundred gold dinars. In the following year, as he was plowing the field, he found a treasure under that same tree and took it to his home. This is to show that whoever does the will of God will be rewarded, as it is said: "The wish of those that fear him he will fulfill" (Psalms 145:19). (See the quotation from the book of Berdichevsky 'Mimakor Yisrael' – from the Sources of Israel, Story No. 653 on the website of Masa, Folk Tale Center of Yoel Perez).

This story appears for the first time in "Midrash Aseret Hadibrot" – Midrash of the Ten Commandments. According to Anat Shapira, who edited the Midrash and published it in a scientific edition (Shapira 2005), this Midrash was written around the 10th or 11th century in Babylonia or Persia. The carob tree is mentioned several times in the Babylonian Talmud, and it seems that our ancestors knew it very well. It is reasonable to suppose that the belief in the demon within the carob tree originates in the Near East and is widely known today in Israel.

On the negative qualities of the carob

Among the Muslims of Israel, we find a number of proverbs that testify to the "negative" qualities of the carob. A slightly scornful attitude toward the carob is shown in the following proverb: "A dirham of sugar in a kantar of wood" (Umm Mahmud, 2.8.04, Ma'ale Tzviya). Literally, it means a small quantity (the dirham is a weight measure of about 3.2 grams) of sugar (the honey of the fruit) in a kantar (a weight measure of about 288 kg) of tree a little sugar in a large quantity of tree. The meaning of the proverb is "a lot of investment with little profit", or as a metaphor for a useless person. The bad name of the carob echoes in many other proverbs. When one wishes to describe days of trouble and adversity, they are compared to the fruit of the carob: "The black days like the horns (fruits) of the carob".

About a useless man, it was said that he was like the pod of the carob which cannot be eaten (more precisely, to be bitten) or to chew it (هو زي قرن الخروب ما بنعضّ ولا بنمضغ,)

(Suleiman Salah Hutaba, 21.12.03, Arraba).

On a man of contradictory qualities, it was said: "He is a man like a carob, sweet and crooked"

(فلان مثل قرن الخرّوب حلو إعوج).

(Sheikh Abu Muhammad Hussein Almaz, 1.11.11, Ka'abiyya).

We will mention here the remark of the Greek priest Philotheos from Acre: "The carob is a symbol of sin, sweet at the beginning and bitter afterwards" (22.3.07).

The fruit of the carob tree is twisted like a horn, and this is the reason for its genus name "*Ceratonia*" derived from the Greek word "Keratos", which means "horn". When one wishes to reprove someone, who is not noted for his good qualities, it is customary to say: "Do not be like the carob fruit which is crooked, be straight" (لا تكُون مِثِلْ الخَرُّوبْ إعْوَجْ)

(Khaled Ali, 11.7.12, Arab al-Aramshe) or in another formulation: "Crooked like a carob"

(إعْوَجْ زَيْ الخَرُّوْبْ).

(Nimer Nimer, 6.12.11, Hurfeish; 'Ataf Suleiman, 19.10.14, Eilabun).

Someone who is not endowed with good qualities is "like a carob" (مثل الخروب)

(Tamir Falah, 11.8.04, Mazarib; Sheikh Abu Muhammad Hussein Almaz, 1.11.11, Ka'biyya).

And everyone will understand what is meant, since a straight carob fruit has never been seen in our district.

"Straight as the fruit of the carob" (Muhammad Ali Jedidat, 12.11.04, Bir Maksur).

The trunk of the carob tree is occasionally hollow and dark, and one can expect that someone is hiding there. This is how a Bedouin who zealously guards his secrets is described: "The secret of the Bedouin is dark like the carob tree"(البدوي سرُّه عِتِم مثل الخروب) (Amin Kozli, 25.2.15, Khawled).

"His heart is dark like the heart of a carob tree" (Muhammad Ali Jedidat, 12.11.04, Bir Maksur).

One legend tells the following story:

> A king had a beautiful daughter. One day the girl was raped, and they blamed it on a young woman who looked like a young man. When they executed the accused woman, her true identity was revealed, and then they regretted the act. Over the grave of the young woman grew the zambak flower (a general name for the iris, the sea-lily, and the Madonna lily, all of them white flowers that testify to the righteousness of the dead). Since the daughter of the king had agreed to the act of rape, they buried her, and a carob tree grew over her grave that was regarded as a bad tree.
>
> (Sheikh Hussein Shahin, a Druze religious leader from Beit Jann, 13.11.05)

A proverb widely known in the villages of the Galilee says that: (بعد شراشِك عالخرّو بة !)

This means: your "ragged" clothes are still hanging on the carob tree. The villagers used to hang their washing on trees. The proverb is used when they want to criticize someone who pretends to have forgotten his recent personal past. The story behind the proverb is about a village girl and her city bridegroom. During her first visit to her parents' home in the village, she asked her father: "What is that crooked fruit hanging on this tree?" Her father answered her angrily and in amazement: "your rags are still hanging on the carob tree" which means: my daughter, you have forgotten your origins too quickly! This proverb is used in connection with anyone who denies or forgets his past, especially his recent past (Saleh 'Aqil Khatib, pers. com.).

The carob tree, king Solomon, and the destruction of the temple

There are many legends about the "bad character" and evil portent of the carob tree. We heard the following story from the 95-year-old Abu Razi from Bu'eina Nujeidat (16.8.04):

> King Solomon used to wake up each morning and see a new tree appearing in the Holy of Holies. The king asked the tree: "What are you doing here?" The fig tree was the first that answered: "I came to bless you". Thus, every morning a different tree appeared to bless him. One day the carob tree appeared. King Solomon asked it: "What are you doing here?" and the carob tree answered: "I am the last of the trees and I have come to destroy the Temple". Four hundred years later, Nebuchadnezzar came and destroyed the entire country.

This story and others like it are widespread in Arabic folklore (Teuma 1987: 84–86) and originate from what is written in the Quran:

> And when we decreed his death, no one revealed to them his death except a worm of the earth that ate his staff; and when the demons saw him fall dead upon the ground they said: Had we known this secret we would not have remained in our punishment and humiliation.
>
> (Surah 34, verse 13)

The following legend is taken from the Talmud:

> At first, King Solomon reigned over the upper world, as it is said: "Then Solomon sat on the throne of the Lord". But finally, he only ruled over his staff, as it is said: "And this was my portion from all of my labor".
>
> (Tractate Sanhedrin 20:2)

Although neither of the two sources indicate what kind of wood was used for the staff of Solomon, the Arabic tradition attributes it to the carob tree based on the similarity between the name of the tree and the verb "*harab*" (both in Arabic and in Hebrew) which means "to destroy" (Teuma 1987: 84). Al-Wahidi, a 5th-century commentator on the Quran who dealt with the story of King Solomon's death, tells the following story:

> King Solomon used to ask every young tree: "What kind of disease can you heal?" and the tree would then answer: "This and that". When the carob tree grew up it was asked: "What is your purpose?" and it answered: "I am destined to destroy your Temple". When the king asked: "Are you really going to destroy it?" the answer was: "Yes!!" Solomon retorted: "What a malicious tree you are!" A short while after this he passed away.
>
> (Wahidi 1989: 9)

Tawfiq Cana'an presents a more complex story which ties all the ends together:

> King Solomon then prayed, uprooted the carob tree, and planted it in his garden. . . . When it became a sturdy plant, he uprooted it and made a walking stick from it. Many years earlier, Belkis, the Queen of Sheba, arrived to pose difficult questions to the king. One of them was how a silk thread could pass through a bead. Solomon called for help from all the animals, birds, creeping creatures, and worms, and only one small white worm carried out the task. It took the end of the thread in its mouth and crawled from one end of the bead to the other side. King Solomon granted the request of the worm to live in any plant that it chose as its food. Without his knowledge, the worm found a place under the bark of a carob tree branch from which the king's bed was made. . . . When the

> day came for the king to die, he was reclining on his bed as usual. Just as the Angel of Death came to take his soul, the worm gnawed into the bed from within it. Suddenly, it broke down, and the body of the king rolled onto the floor.
>
> (Canaan 1927: 37–38)

Fairy of the carob tree

In their book, Crowfoot and Baldensperger (1932: 113) give concrete descriptions of dangerous creatures that exist in the carob tree:

> A woodcutter in Ramallah set out to work one morning and met a bride covered in gold emerging from a carob tree. She spoke very sweetly to him and promised him great wealth if he threw away the axe on his shoulder and followed her. When he refused to comply with her request, she struck at his eyes and blinded him.

Conclusions

In spite of the dispute whether or not the carob had existed in Israel since biblical times and when it arrived there (Liphschitz 1987; Kislev 1988), it is accepted today that the carob is a tree of Middle Eastern origins (Ramón-Laca and Mabberly 2004; Zohary 2002), which may have been less widespread in earlier periods (Liphschitz 1987; Kislev 1988). According to quotations from the Talmud, it appears that it was considered a cultivated tree (Felix 1994: 203–214). It is not surprising that this gave it positive perception, which reached its peak among the Jews during the Talmudic period. The tree was highly regarded in Arab villages, mainly because of the concentrate syrup ("*rhub*") of its fruit, which was an important source for sweetness known as "carob honey". It is more difficult to explain the "curse" that overhangs the carob as the abode of demons, at least since the 10th century CE. This negative attitude to the carob as an accursed tree and a place where demons reside has echoes in Jewish and Muslim legends that link the carob tree with King Solomon and the Destruction of the Temple by the Babylonians. These traditions are also potent and pervasive today throughout the country. The link between the carob tree and snakes seems to be peculiar to the folklore of this country, both among Muslims and among the Druze.

Contemporary Arab proverbs connected to the carob are all negative in character, referring to the crooked-shaped fruit and the "dark" heart of the tree. In old carob trees, the trunk is hollow and often of a blackish color. In addition to the darkness within it, it appears to be threatening and a haven for snakes. The claim that the black color of the fruit is the source for its bad name seems tenuous in comparison with the great importance of the carob, mainly as a source of food for the poor and in times of famine. This has surely caused some measure of "injustice" to its good name.

Our results from contemporary interviews show that the Carob tree has a good spiritual reputation among Jews and Druze while most of the negative attitudes were expressed in the Muslim sector. It is quite a puzzling finding, since the Carob is an important plant in the local Muslim domestic use as an edible plant and as well as a source for the highly praised sweetening syrup.

References

Altschuler, D. 1776. *Ayalah Shluha – Yesha'ayahu (Isaia)*. (In Hebrew). Available at www.hebrewbooks.org/58360

Canaan, Tewfik. 1927. *Mohammedan Saints and Sanctuaries in Palestine*. Luzac and Co., London.

Chandran, M. D. Subash, and J. Donald Hughes. 1997. The Sacred Groves of South India: Ecology, Traditional Communities and Religious Change. *Social Compass* 44(3): 413–427.

Crowfoot, G. M. H., and L. Baldensperger. 1932. *From Cedar to Hyssop*. Sheldon Press, New York and Toronto.

Dafni, A. 2006. On the Typology and the Worship Status of Sacred Trees with a Special Reference to the Middle East. *Journal of Ethnobiology and Ethnomedicine* 2(1). https://link.springer.com/article/10.1186/1746-4269-2-26.

Dafni, A. 2010. *Sacred Trees of Israel*. Hakibutz Hameuchad, Tel Aviv. (In Hebrew).

Dafni, A. 2011. On the Present-day Veneration of Sacred Trees in the Holy Land. *Folklore: Electronic Journal of Folklore* 48 (2011): 7–30. www.researchgate.net/profile/Amots_Dafni/publication/262661017_On_the_Present-day_Veneration_of_Sacred_Trees_in_the_Holy_Land/links/549169990cf2d1800d88366f/On-the-Present-day-Veneration-of-Sacred-Trees-in-the-Holy-Land.pdf.

Dafni, A., and S. A. Khatib. 2017. *Plants, Demons and Wonders: Plants Folklore of the Holy Land*. Olam Chadash, Tel Aviv. (in Hebrew)

Eliade, M. 1958. *Patterns in Comparative Religion*. Sheed & Ward, London.

Felix, J. 1994. *Fruit Trees in The Bible and Talmudic Literature*. Rubin Mass Press, Jerusalem.

Frazer, J. G. 1923. *Folklore in the Old Testament*. Tudor Publishing Company, New York.

Hadad, M. 1981. The Folklore of the Four Communities in Peqi'in. *In Proceedings of the World Congress of Jewish Studies*, pp. 129–137. World Union of Jewish Studies (In Hebrew), Jerusalem, Israel.

Hamilton, L. S. 2002. Forest and Tree Conservation through Metaphysical Constraints. *The George Wright Forum* 19: 57–78.

Hehn, V. 1885. *Cultivated Plants and Domesticated Animals in their Migration from Asia to Europe: Historico-Linguistic Studies*. Sonnenschein, London.

Howell, D. 1970. Health Rituals at a Lebanese Shrine. *Middle Eastern Studies* 6: 179–188. www.daat.ac.il/daat/kitveyet/emunat/10/01004.htm.

Kislev, M. 1988. The History of the Carob in Our Land. *Halamish* 6: 20–30. (In Hebrew).

Liphschitz, N. 1987. *Ceratonia siliqua* in Israel: An Ancient Element or a Newcomer? *Israel Journal of Botany* 36: 191–197.

Malossini, Andrea. 2013. *Italian Superstitions. Body, Mind & Spirit*. Milan, Italy.

Millar, D., Richard A., and P. Atoyure. 1999. Shrines and Groves. *COMPAS Newsletter*, October 1999, pp. 32–36. Available at www.compasnet.org/blog/wp-content/uploads/2011/03/CM/CM2.pdf. Accessed on July 30, 2011.

Mills, J. 1864. *Three Months' Residence at Nablus and an Account of the Modern Samaritans*. J. Murray, London.

Philpot, J. H. 1897. *The Sacred Tree or the Tree in Religion and Myth*. MacMillan, London.
Pliny 1945. *Natural History*, translated by H. Rackham. Harvard University Press, Loeb Classical Library, London & Cambridge, MA.
Porteous, A. 1928. *Forest Folklore, Mythology and Romance*. George Allen & Unwin Ltd, London.
Quantz, J. O. 1898. Dendro-psychosis. *American Journal of Psychology* 9(4): 449–506.
Ramón-Laca, L., and Mabberley, D. J. 2004. The Ecological Status of the Carob-Tree (*Ceratonia siliqua*, Leguminosae) in the Mediterranean. *Botanical Journal of the Linnean Society* 144: 431–436.
Robertson-Smith, W. 1889. *Lectures on the Religion of the Semites*. Adam & Charles Black, Edinburgh.
Schwartz, T. 1995. *Carob the Jewish Tree*. (In Hebrew). Available at www.daat.ac.il/daat/kitveyet/emunat/10/01004.htm
Shapira, A. 2005. *Midrash aseret ha-dibrot: text, mekorot u-ferush* [Midrash of the Ten Commandments: Text, Sources and Interpretation]. Mossad Bilaik, Jerusalem. (In Hebrew).
S. J. M. Gray. 1995. Trees. In *The Encyclopaedia of Religion*, edited by M. Eliade, vol. 15. Macmillan, New York, pp. 26–33.
Skinner, C. M. 1913. *Myths and Legends of Flowers, Trees, Fruits, and Plants in All Ages and in All Climes*. JB Lippincott Company, Philadelphia, PA.
Teuma, E. 1987. The Solomon Legend in Muslim Tradition. *Melita Theologica* 38: 81–87.
Wahidi, A. A. 1989. *Asbab al-Nuzul. Beirut: Maktabah al-Tsaqafah. Ahmad al-Wahidi, Asbab al-Nuzu*. Maktabah al-Tsaqafah, Beirut. (In Arabic).
Zahan, Dominique. 1979. *The Religion, Spirituality and Thought of Traditional Africa*. Chicago University Press, Chicago.
Zohary, D. 2002. Domestication of the Carob (*Ceratonia siliqua* L.). *Israel Journal of Plant Sciences* 50(suppl.): 141–145.

10 Plant pathogens in emotional landscapes

Olive stakeholders and the *Xylella fastidiosa* outbreak in Apulia, Southern Italy

Fabio Gatti

Introduction

> I kept going because mine is a family farm, I had an emotional bond with it . . . and you can't run a business with your heart . . . you have to manage it with your accounts. We handled it with our hearts, and this put us in trouble. . . . If I had stopped ten years ago, everything would be fine now.
>
> (an olive farmer, from interview)

"Look, it's a disaster". Giuseppe points at some of the olive trees located in a field adjacent to the road, while we drive toward one of his olive orchards in the surroundings of Oria, a small town in the southern Italian region of Apulia. From having a quick look at them, they don't seem to have the ordinary luxuriant and verdant appearance with which olive trees usually look back at distracted visitors like me all year round: with some of the branches completely withered, while others still in vegetation, they show an unusual spotty leopard dress, which they have never worn in their whole centuries-old lives. "We can only hope that they will find a treatment . . . other than that now it's too late, it's not gonna stop".

Giuseppe is an olive farmer from three generations. In his 60s now, he owns almost 100 hectares of olive tree orchards distributed over the districts of Lecce, Brindisi and Taranto, the southernmost districts of the region. "I really don't know what to do", he says while telling me his story. What we are discussing is something that for many olive farmers of the region, has become the daily source of trouble and concern for some years now: the symptoms of a disease called *Complesso del Disseccamento Rapido dell'Olivo* (CoDiRO) – Olive Quick Decline Syndrome (OQDS) in its English translation. CoDiRO started to manifest itself around 2010 in the countryside of Gallipoli, on the southwestern coast of the Lecce district, and spread unexpectedly fast over the territory, causing a massive change in what was, for a local like me, the familiar landscape. In a very short time, over vast areas of the heel of the Italian peninsula, the typical robust, magnificent and evergreen lush-foliage silhouettes of centenary olive trees disappeared and gave way to shrunken and skinny, dying trunks:

DOI: 10.4324/9780429320897-12

Figure 10.1 A desiccated tree landscape in the countryside of Ugento, November 2017.
Source: (Photo: Janos Chialá, www.postphotography.eu)

clearings of dead bodies extending over many and many kilometers hit the eye of any human walking across the area (see Figure 10.1).

While the epidemic has been mostly framed as a technical agricultural problem, one in which an alien bacteria known as *Xylella fastidiosa*, coming from outside the European borders, found the perfect playground for its proliferation, this contribution attempts to look at the socio-anthropological features and emotional consequences of the outbreak, re-framing the debate in its more political terms. What does such an agricultural emergency imply for the whole agricultural model, and how does that affect the subjectivities, emotions and identities of the local people in the area? And in what way did these subjectivities, emotions and identities feed into resistance and resilience to this challenge? Specifically, to what extent does the resistance to the eradication of olive trees (the main containment measure enforced by European, national and regional authorities) represent a different way of looking at nature and at the interaction between human and non-human subjects?

As will be clear, the case reveals the coexistence of different ways of valuing the olive tree as an entity, different world-views that in Apulia have been coexisting for a long time but now experience incompatible tensions and are starting to clash due to the sensation of loss caused by the "blasted landscapes" (Tsing 2014) generated by such "feral biologies" (Tsing 2015). On the one side, there is the techno-modernist agricultural paradigm aimed at fostering an intensive, highly mechanized agrifood system in the belief of that being the only way for agriculture to increase productivity and efficiency, and in this way be able to feed an increasingly populated planet. On the other, there is an agroecological paradigm in which olive trees are more than just an economic asset to be exploited for making profits, they are sometimes "part of the family", living beings that "are asking for help" and that farmers have the responsibility to care for, because on this depends our own well-being.

The chapter builds upon the broader strand of literature on other-than-economic ways to value nature (Rolston 1981; Martínez-Alier 2002) and on how management of natural resources and ecosystems is a complex mix of economic considerations together with non-economic (but no less valuable) factors, often related to deep symbolic meanings attached to the natural world. In Apulia, for example, the connection among the rural population and olive trees is so strong that the threat posed by *Xylella fastidiosa* must be equally read as a threat to the history, the identity and the imagination of a whole community. In this context, the extra economic value of olive trees cannot be overlooked, and it has to be acknowledged as being firmly at the core of the resistance attitude toward the disease management measures enforced over olive orchard owners.

Additionally, my aim is to briefly engage with the so-called "emotional turn" in the social sciences (Massumi 1995; Anderson and Smith 2001; Lemmings and Brooks 2014) and contend that love and affection toward the olive trees is an essential component to be considered when trying to understand the different ways in which olive stakeholders dealt with the management of the disease. Because of this, the sense of loss due to the death of the olive trees and the emotional bond that drives the different forms of resistance to the radical transformation of the landscapes – together with the memories and the subjective meanings that accompany them – play a central role in this research.

The chapter is structured as follows. First, I introduce the case and outline the genealogy of the epidemic starting from the detection of the bacteria in 2013. Second, I will engage with the feelings of loss and anger that characterize the area where most of the desiccated trees are located, namely the infected area. Third, I will look at the role that deep socio-cultural meanings play in the resistance to the challenge of the *Xylella fastidiosa* epidemic in the containment area, and how this resistance represents, in fact, a different view over the future of agriculture in the area. I will then provide some concluding remarks.

Context and case study

With about 60 million olive trees and 277,000 hectares of land cultivated with olive trees, representing 32% of the total national amount (Ismea Report 2018), Apulia is one of the most important olive groves sites in the whole of Europe. Within these 60 million, about 5 million are estimated to be millenarian: their trunk circumferences can reach eight to ten meters and assume the most disparate shapes. In light of such majestic morphologies, some of them are designated as 'monumental olive trees' (*ulivi monumentali*) and "by virtue of their productive function, of ecological and hydro-geological defense, as well as the peculiar and characteristic elements of the regional history, culture and landscape", they are protected by the Apulian regional legislation. Most of them, in many parts of the Apulian territory, give shape to the extended monocultural landscapes distributed over the territory.

The history of olive monoculture in southern Italy originates in 1739, when Charles III of Bourbon introduced specific tax exemption measures for the planting of olive trees, in order to foster the production of olive oil meant for export to the Northern European markets where the machines made possible by the industrial revolution were starting to enter the stage (Bandiera 2019: 46). At that time, olive oil was used mainly as a raw material for the soap industry or as a lubricant for those machines, and this is the reason why Salento is known for producing *olio lampante* (lamp oil), a low quality, inedible olive oil obtained by olives collected from the ground in an advanced state of maturation. This process of subsidizing farmers for planting olive trees is thus considered key for initiating the modern olive cultivation practice with monocultural features, and therefore crucial in shaping Apulian landscapes in the form they look today (Bandiera 2019).

It is in this context that the Xylella epidemic exploded. At the beginning of 2010, some of the olive trees around the area of the city of Gallipoli, on the southeastern coast of the region, started to show manifestations of a sickness consisting of the "withering and desiccation of scattered terminal shoots, which rapidly expands to the rest of the canopy and results in the collapse and death of the tree" (Loconsole et al. 2014: 7). Initially described as a "severe olive decline of unknown etiology" (Nigro et al. 2013: 668), the disease was called Complesso del Disseccamento Rapido dell'Olivo (CoDiRO) or, in its English version, Olive Quick Decline Syndrome (OQDS) (Colella et al. 2019). Possible causes were considered to be the insect *Zeuzera Pyrina*, some different varieties of fungus such as *Phaeoacremonium*, *Phaemoniella*, *Pleurostomophora* and *Neofusicoccum*, as well as the bacteria *Xylella fastidiosa* (Nigro et al. 2013; Saponari et al. 2013).

Xylella fastidiosa is a quarantine pathogen classified as high risk by the European Food Safety Authority (EFSA) (EFSA 2015). It colonizes the xylem of the host plants and gets transported from plant to plant via different spittlebug insect vectors that, after having fed themselves from the same xylem, bring it from contaminated individuals to bacteria-free ones. Historically confined to the Americas (mainly Brazil, Costa Rica and southern California, but also present in Argentina and Mexico), where it is endemic and associated with numerous diseases such as Pierce's disease in vineyards and various leaf scorch diseases affecting citrus and almond plantations, its unexpected detection on olive trees in Apulia in 2013 sparked immediate reaction of the phytosanitary authorities. With Decision 2015/789, the European Commission deliberated on the emergency measures to be implemented in order to limit the further spread of the pathogen in the EU territory (Pagano 2017). The decision implied "establishing a demarcated area around infected areas with specific requirements associated with surveillance, plant removal and other management measures including agricultural practices to control vector populations" (EFSA 2019: 3).

The most disputed measure was the "plant removal" one, namely the injunction of eradicating all trees which were found infected by the bacteria, plus all surrounding trees, regardless of their health status, within a radius of 100 meters around the infected plants. The reaction from environmental activists, agro-ecological farmers and local civic associations was immediate: ranging from attitudes of denial ("Xylella does not exist") to the belief that "you don't die of Xylella", the idea that the only way to deal with the disease was the eradication of centuries-old trees was just unacceptable and inconceivable for most olive stakeholders. Some people bound themselves to trees in a desperate attempt to block plant uprooting, others engaged with a permanent garrison in order to stop the arrival of bulldozers designated for the eradication. The question that thus arises is the following: why did all this happen? What motivated such a reaction to the eradication of ancient olive trees in Apulia? In order to understand it, we have to dig a bit more into the intimate relations between Apulian inhabitants, olive trees and their most valuable derivative product: olive oil.

Human-plant relations in the context of *Xylella fastidiosa* epidemic

In their wild state, olive trees (*Olea europea*) are shrubs. They have been growing spontaneously in the warm areas of the Mediterranean since pre-Roman times (Primavera et al. 2017) and present characteristics which are quite different from the "domestic olive tree". Besides the shrubby habit, young branches (the ones which start from the base of the tree) are hard and spiky, leaves are smaller and have a more rounded shape compared to its domesticated relative, and fruits are smaller and with a bigger stone. Transition from the wild variety to the domesticated one occurred over the years through genetic selection derived from various changes in climate as well as human selection of the plants with larger fruits and better features for human use.

More than the olive tree alone, what is also extremely important in Apulia is the result of its cultivation: olive oil. This is not the place for describing the richness of properties of olive oil, nor for presenting in detail the art of olive oil making. What matters here is to stress how important olive oil is in Apulian culture: every year, thousands of people gather in the countryside to collect olives and bring them to the press. Not all of them make a living out of olive farming and olive oil production. Many of them, including me, just make the olive oil they need for family consumption, from pieces of land belonging to the family and inherited over the years. It is a yearly celebration: at the end of the harvest, it is very common to see families and friends gathering around a meal and tasting the first, rigorously raw, olive oil of the season. It is a ritual which keeps the memory of the family alive and is a collective act of remembrance.

Olive oil, and the olive trees that contribute to co-produce it, sometimes serve as an intergenerational link: as some peasants in Apulia used to say, "you do not plant olive trees for yourself. You plant them for your sons, and for those

Figure 10.2 Me at the age of four, photographed next to some recently planted olive trees in the family field.

who will come after them" (Chialá 2019: 69). In Figure 10.2, you can see a young me (about 4 years old) next to some olive trees which my father planted when I was born. The reason behind this research can be probably summarized with this picture. Every time I go back to visit my house, I take a walk between the trees, look at them and pay attention to their health status. Even now while writing, I am in the middle of the harvesting season, the time in which, together with my father and few other friends, we collect the fruits and bring them to the mill to make our own, family consumption olive oil.

The threat of the *Xylella fastidiosa* epidemic has been affecting me, as well as many other olive orchard owners, creating anxiety and fear: what if in a few years I won't have any olive trees *alive* anymore in my little orchard? What if everything around my little house will turn into the landscape in Figure 10.1, just a cemetery of skeletons surrounded by fog? It was not uncommon, in the past harvest season to think: "It might very well be that in a couple of years these trees will be all dead, so let's make the best out of this now". Even if I don't make a living out of olive oil production, the loss of such heritage would be tremendously hard to deal with for me and my family. And I believe it would be the same for most of the olive farmers around the area. Whether you are a big farmer owning tens of hectares, or a small farmer devoted mainly to family production, there is a thin thread that connects all different stakeholders: the sense of loss and grief generated by such devastated landscapes. It is to this sense of loss and emotional pain that I will turn in the next section.

Blasted emotional landscapes

From this section on, my aim is to share with the reader some of the emotions I have experienced and encountered among olive farmers during my fieldwork among Apulian olive farmers in August and September 2019, and in October and November 2020. While not pretending to engage with the debate about the distinction between affects and emotions in the social sciences (Pile 2010), what I want to do here is to describe some of the affective encounters that I have witnessed between olive farmers and olive trees. Affective encounters are what Archambault (2016: 249) describes as "a meeting with someone or with something [that] triggers some sort of effect; when it inspires, unsettles, troubles, moves, arouses, motivates, and/or impresses". In the *Xylella fastidiosa* epidemic case, such an encounter between people and olive trees has motivated, for example, the resistance against the eradication of the trees.

"Affective encounters are shaped by the particular socio-historical geographies within which they are embedded" (Archambault 2016: 250), and the history of olive plantations is such that every Apulian inhabitant is embedded in a vast simplified monocultural landscape that has become, slowly, the only way we could imagine Apulian landscapes. In the words of one researcher from the National Research Council (CNR) in Bari I interviewed:

> We have always considered olive trees eternal. Now, this small being is destroying not only olive trees, but also our conception of olive trees. . . . *Xylella fastidiosa* is not only harmful for olive trees, it is harmful also for the people: the olive tree as the reference point has disappeared. You lose production, but you also lose a point of reference: this is what secular olive trees were for Apulian farmers.

A clear spatial distinction must be introduced here. In fact, following the EU resolution, in 2015, the Italian government implemented a containment strategy based on the division of the Apulian territory into different demarcated areas: an infected zone (*zona infetta*) where the presence of the bacteria has been verified by molecular investigation methods but no compulsory measures (either plant removal or pesticide use) apply; a containment area (*zona di contenimento*) that includes the last infected 20 km up to the point where the last infected tree has been found, and where most of the monitoring and eradication efforts are concentrated; and a buffer zone (*zona cuscinetto*), a strip of land extending 10 km north of the containment zone and representing the start of the disease-free portion of the territory, and where surveillance is meant to monitor the spread rate of the epidemic. As soon as one tree is found infected in the buffer zone, the containment zone moves up to that point, and the buffer zone shifts accordingly.

These different demarcated areas also correspond to different emotions with respect to the epidemic: in the infected zone, a sense of abandonment and powerlessness prevails, due to the fact that after so many years "farmers have been left alone" to fight against the blight. It is also the area where "blasted landscapes" of Figure 10.1 are by now the daily bread, and where debates are already shifting from "how to fight the bacteria" to "how to convert a territory devoted to olive production to something else". Conversely, in the containment zone feelings of hope and "certainty that I could make it", partially fostered by the fact that the disease is (still) not very visible and that the number of plants showing symptoms of desiccation is still quite low, are the dominant ones.

Anger and hopelessness in the infected area

The area where the feeling of loss is most pronounced is the infected area, where most olive trees are infected and the typical landscape is the one shown in Figure 10.1. As previously mentioned, here, the sense of loss is predominant among the rural population, and it gives rise to what Cunsolo and Landman (2017) call "ecological grief". Some civic associations have even staged a funeral for the olive trees. What is most compelling is the feelings of abandonment and hopelessness that I have encountered among most of the farmers. One of them, when asked if he believes whether there can be some treatment that helps the tree to survive the infection, or a cure for the disease, told me: "Don't listen. . . . Those are all bullshits! There is nothing you can do, nobody talks about it anymore. You die, that's it. . . . You just die."

Giusy, a female farmer owning around 100 hectares of olive trees in the small village of Giuggianello, about 560 km south, shares similar, if not more desperate, emotions. A third generation farmer, in 2008 she and her co-owners, one of her brothers and her cousin, decided to restructure the family farm betting everything on olive farming. They applied for a mortgage of 450,000 euros and bought the machines and the shed for building a new olive mill. In July this year, they had to sell everything "for an insanely cheap price" to a Tunisian buyer, after having sold year by year pieces of land in order to be able to meet the mortgage payment deadlines. Land which, in the same way as their milling machines, had seen a decrease in market value since the start of the outbreak: "before Xylella, one hectare of land was valued twelve, fifteen thousand euros. We sold it for four or five thousand. It's a sacrilege, especially for the ones who cultivated the land legally and with their own efforts".

She welcomes me in a sumptuous but apparently decaying house, as a witness to a glorious past that no longer has the strength to sustain itself. Her character is modest, there is a kind of frustration and anger, including a bit of hysteria, in her looks and movements. I have a feeling that she expects me to be able to do a lot more than I am actually capable of. She tells me when I ask to tell me her story,

> We had a considerable extension of the land, from three generations . . . but now we are dead. In other words, *there is no hope*. All the farms in the area, the normal ones, are all cut short . . . there is no solution.

Her link with the family is clear from each of her words. She explains,

> When I was young, I used to cycle among the trees, in our land. I loved to cycle, and I went there very often. I was calm and peaceful as that was our land and nothing could happen to me. Also when I grew up, I kept

> cycling among them. Now, it's been two years that I don't touch my bike: it's full of rust. . . . I can't turn around in the midst of olive groves that are all dried up. . . . It hurts too much to look back and remember what it was: the family, the farm.

Anger is one of the emotions that I frequently detected among farmers. It is an anger that takes multiple forms and has multiple recipients, a matrix of anger that needs to be carefully unpacked. There is the anger toward the authorities who want to introduce intensive and superintensive plantations in a territory where there is a chronic lack of water resources, but also the anger toward the people who opposed themselves to the eradication "because in this way they have helped spread the bacteria" and therefore undermined the efforts of hundreds of big farmers who are now on the verge of bankruptcy. There is the anger toward those farmers keeping their orchards in a state of abandonment (e.g., not clearing the weed and therefore indirectly increasing the population density of the insect vector responsible for the contagion) but also anger toward politicians that with decades of failed policies "have decided to kill an entire sector slowly, not with Xylella". There is the anger toward the 'madmen' who talk of immune defenses of olive trees undermined by reckless use of pesticides and herbicides and that, according to some researchers, "does not make any sense"; and the anger toward the agricultural experts who have most of the time denigrated and belittled any form of lay knowledge and empirical expertise by farmers.

Salvatore, a small farmer in the town of Salve, gives me, however, a more positive feeling. He also experiences a sense of anger and powerlessness but claims that he wants to save his olive trees "cost what it may". Still, he feels alone in this arduous undertaking:

> These plants are asking me for help and I feel helpless. I am alone. It's like if my son was falling into a ravine, and he . . . climbs, clings with his nails . . . and I can't . . . I'm there a few inches from him . . . and I can't help him, save him.

He believes that everything has to do with a broken balance, something that he and most of the farmers of the area have been contributing to thanks to "bad recommendations by agronomists". For him, it is now time to reconsider the whole agricultural model, abandon monoculture and "recompose the balance [by restoring] organic matter and microorganisms". In line with the latest advancements of regenerative and restorative agriculture, he believes that "soil is the intestine of the planet" and that we have to restore biodiversity if we want the ecosystem to be able to cope with other living beings that we don't have to eradicate, but rather learn to co-live with.

This position is also shared by agronomist and plant pathology researcher Marco Scortichini, a bacteriologist from the Research Centre for Olive, Fruit and Tree Crops (CREA) that has developed a protocol that has proven able to let the olive tree stay productive, even though it doesn't represent a definitive cure for the disease (Scortichini et al. 2018). According to him,

> Eradication measures make sense when you are talking of a limited extension of the outbreak. In the case of trees, 1–2 hectares, maybe 10, maybe 50. But not when you have 8000, 10000 hectares infected, moreover of a pathogen that doesn't affect olive trees alone. So the approach has been since the beginning extremely impractical.

He also stresses the fact that olive trees are not potatoes, tomatoes or other species that you can destroy one year and immediately replant the year after. "Olive trees have an incredibly ancient history, so you cannot conceive a strategy to solve the crisis like that".

This brings us to the original argument of the chapter: olive trees have such a strong cultural value in Apulian society, that they cannot be treated like any other plant species. In some cases, they are even talked about as someone who is part of the family: "They are like a newborn baby to me", Salvatore told me when asked what olive trees are for him. And as every newborn baby has a special relation with his or her mother, he claims to have a special relation with his trees. He "treats [his] land and [his] plants like [his] children", and he spends so much time in the field with his trees that sometimes his wife "is not jealous of women but of plants". He firmly believes in the need for spending time with his plants in order to be able to understand them:

> If that plant is sick . . . if I'm always in the fields I will see it, I will look at it, and I will understand it. But if I go to the fields only when I have to harvest, and I'm always in a hurry . . . I will stress [the olive trees]! Trees communicate! I communicate with the trees. And between them they also communicate, they know if I am there in order to help them, or if I'm going to exploit them.

Trees are "like elderly people, they need company", and therefore sometimes "I go to the countryside, I put myself a little under that tree, a little under the other, and we talk".

Love and resistance in the containment area

> I always tell my children: I give you the olive grove, but you have to love it!

In the containment area, the landscape is quite radically different: no sign of desiccation can be spotted at first sight, and florid tree crowns smile at me when I visit the site. In 2018, in the town of Cisternino, Franca and other activists of the environmental association COSATE – Comitato per la Salvaguardia dell'Ambiente e del Territorio della Valle d'Itria – have been carrying out a "permanent defense" (*presidio permanente*) around an olive tree that regional authorities wanted to eradicate after having assessed the presence of the bacteria. "They came at night, police blocked the access roads to the site and forbade the person who was here to record anything. When we arrived, the tree had already been cut", she told me when we met at the "crime scene".

She explained to me that "the olive tree is a tree that has lived for a long time because it lived in symbiosis with the peasants who respected their status . . . so they had placed them far from each other, cleaned them, pruned them". Abandonment and lack of care is thus one of the factors that contributed to the disease. Now, "they want to replace peasant agriculture with industrial-type agriculture, that is, done with mechanical means". And in order to do this, "it doesn't matter if you [cut] these trees, which can be five hundred, a thousand, a thousand and five years old". It is evident from her words that what is at stake here is not only the survival of a species, the olive tree, that has extraordinary extra-economic value. It is a whole agricultural model that is put under the spotlight. According to most agricultural experts, traditional olive farming, due to its low productivity, cannot compete with highly mechanized olive production landscapes such as the ones that can be found in Spain, or with non-European olive producers such as Morocco and Tunisia, with their much lower cost of labor.

In the countryside of Ostuni, Vitantonio confirms this: "the big problem is that in our territory we have enormous production costs". He is aware of what happened in Salento ("your heart cries, it became like a desert") but his position on the eradication measure is interestingly not as strict as the positions of some other farmers I had the opportunity to talk to. In a similar way as Salvatore, he believes that care and constant presence on the field is key for prevention:

> In the past, when you had a desiccated branch, farmers were immediately cutting off the branch. . . . If they saw that the tree was a little withered, they would throw some earth on it, put some manure on it, plow a little more. . . . They were looking after it! Then, if nothing was working, they would cut it down.

Therefore, removing some ill tree is not something unconceivable for him:

> If you come to our olive grove I will show you that we have at least 10 olive trees that are not the original ones. That is, it is not the original plant . . . over time, they saw that the plant did not grow well, they cut it and [planted a new one].

Similarly to other stakeholders, he commented upon the fact that, following the wrong suggestions of academic and agronomic research,

> We [farmers] have insisted that olive growing [must] be done intensively and according to monoculture. Now, are we sure that monoculture and intensive plantations are good for the olive tree? . . . We know that an olive tree is much more exposed [to pathogens] if, for example, it is not pruned, or if it is very narrow. That is, if it is not ventilated: humidity is created inside, and the attack by certain pathogens is facilitated. But this is exactly what you have when you place trees 1 meter apart from each other. So, we

> have made intensive olive plantations, [but] are we sure that we want the good of the olive tree or have we just chased after economic profit?

Once again, it's the whole profit-based monocultural model to be put in the dock. Nevertheless, replanting schemes from regional authorities subsidize farmers whose plantations have been affected by *Xylella fastidiosa* to replant tolerant and resistant varieties, only if irrigated and in intensive or superintensive plantations (namely from 1,000 to over 7,000 trees per hectare). According to a retired manager working in the olive oil sector for years "It is madness, that is, a real madness. We don't have water!". Vito repeatedly points to this tension between economic sustainability (i.e., productivity) and the need to preserve a heritage that was left to us by our ancestors: "here we need someone to tell us how to make the millenary olive grove survive, not how to have high productivity". It is, once again, the core of the conflict around the *Xylella fastidiosa* epidemic management policies: shall we take this agricultural emergency as an opportunity to rethink our agricultural model toward a more biodiverse, less intensive and more *environmentally* sustainable food system, or as an excuse to implement a shock doctrine (Klein 2007) to make "shocked societies give up things they would otherwise fiercely protect" (ibid:17)? This is what is at stake in the *Xylella fastidiosa* epidemic in Apulia.

Conclusion

In conclusion, I have briefly outlined an emotional geography of the *Xylella fastidiosa* epidemic in Apulia. Subjective emotions and ecological grief generated by the sudden change of landscapes and the natural environment are in fact essential components of the opposition to the eradication strategy enforced by phytosanitary authorities. Additionally, the contestation of the institutional management strategy has to do with different ways to imagine the future of Apulian agriculture in a post-Xylella setting. Two different paradigms are fighting each other in the face of the outbreak: an agricultural modernization paradigm aimed at fostering intensive-type plantations able to compete on the global markets, and an agroecological paradigm where traditions and local knowledge should be preserved in the context of an agriculture that is respectful of plants, people and the environment. What the result of this fight will be it is still too early to say.

References

Anderson, K., and S. Smith. 2001. Editorial: Emotional Geographies. *Transactions, Institute of British Geographers NS* 26: 7–10.

Archambault, J. S. 2016. Taking Love Seriously in Human-Plant Relations in Mozambique: Toward an Anthropology of Affective Encounters. *Cultural Anthropology* 31(2): 244–271.

Bandiera, M. 2019. Tra Bosco e Monocoltura: la Xylella fastidiosa e le "buone pratiche. *Rivista Epidemia* n.1/2019. Autoproduced Publishing. Available at www.collettivoepidemia.org/.

Chialá, J. 2019. Dopo la Xylella: il dibattito sul futuro dell'olivicoltura in Puglia. *Rivista Epidemia* n.1/2019. Autoproduced Publishing. Available at www.collettivoepidemia.org/.

Colella, C., R. Carradore, and A. Cerroni. 2019. Problem Setting and Problem Solving in the Case of Olive Quick Decline Syndrome in Apulia, Italy: A Sociological Approach. *Phytopathology* 109(2): 187–199.

Cunsolo, A., and K. Landman, eds. 2017. *Mourning Nature: Hope at the Heart of Ecological Loss and Grief.* McGill-Queen University Press, Montreal, QC.

EFSA. 2015. Scientific Opinion on the Risks to Plant Health Posed by Xylella fastidiosa in the EU Territory, with the Identification and Evaluation of Risk Reduction Options. EFSA Panel on Plant Health. *EFSA Journal* 13(1): 3989.

EFSA. 2019. Update of the Scientific Opinion on the Risks to Plant Health Posed by Xylella fastidiosa in the EU Territory, with the Identification and Evaluation of Risk Reduction Options. EFSA Panel on Plant Health. *EFSA Journal* 17(5): 5665.

Ismea Report. 2018. *Olive Oil Production in Italy, Production Data 2017–2018*. Available at www.regione.puglia.it/documents/3652161/5611645/Produzione_2017_18+%282%29.pdf/c60bdcca-79b1-4268-aa6e-1a55a1d07961. Accessed January 14, 2019.

Lemmings, D., and A. Brooks 2014. The Emotional Turn in the Humanities and Social Sciences. In *Emotions and Social Change. Historical and Sociological Perspectives*, edited by D. Lemmings and A. Brooks, pp. 3–18. Routledge, New York, NY.

Loconsole, G., O. Potere, D. Boscia, G. Altamura, K. Djelouah, T. Elbeaino, D. Frasheri, D. Lorusso, F. Palmisano, P. Pollastro, M. R. Silletti, N. Trisciuzzi, F. Valentini, V. Savino, and M. Saponari. 2014. Detection of Xylella fastidiosa in Olive Trees by Serological and Molecular Methods. *Journal of Plant Pathology* 96: 7–14.

Martínez-Alier, J. 2002. *The Environmentalism of the Poor: A Study of Ecological Conflicts and Valuation*. Edward Elgar, Cheltenham.

Massumi, B. 1995. The Autonomy of Affect. *Cultural Critique* 31: 83–109.

Nigro, F., D. Boscia, I. Antelmi, and A. Ippolito. 2013. Fungal Species Associated with a Severe Decline of Olive in Southern Italy. *Journal of Plant Pathology* 95: 668.

Pagano, M. 2017. The Italian Xylella Case: The Role of EFSA in the EU Decision-making on Risk. *European Journal of Risk Regulation* 8: 599–605.

Pile, S. 2010. Emotions and Affect in Recent Human Geography. *Transactions of the Institute of British Geographers* 35(1): 5–20.

Primavera, M., C. D'Oronzo, I. M. Muntoni, F. Radina, and G. Fiorentino. 2017. Environment, Crops and Harvesting Strategies during the II Millennium BC: Resilience and Adaptation in Socio-Economic Systems of Bronze Age Communities in Apulia (SE Italy). *Quaternary International* 436: 83–95.

Rolston, H. 1981. Values in Nature. *Environmental Ethics* 3: 113–128.

Saponari, M., D. Boscia, F. Nigro, and G. P. Martelli. 2013. Identification of DNA Sequences Related to Xylella fastidiosa in Oleander, Almond and Olive Trees Exhibiting Leaf Scorch Symptoms in Apulia (Southern Italy). *Journal of Plant Pathology* 95: 668.

Scortichini, M., J. Chen, M. De Caroli, G. Dalessandro. N. Pucci, V. Modesti, A. L'Aurora, M. Petriccione, L. Zampella, F. Mastrobuoni, D. Migoni, L. Del Coco, C. R. Girelli, F. Piacente, F. Piacente, N. Cristella, P. Marangi, F. Laddomada, M. Di Cesare, G. Cesari, F. P. Fanizzi, and S. Loreti. 2018. A Zinc, Copper and Citric Acid Biocomplex Shows Promise for Control of Xylella fastidiosa subsp. pauca in Olive Trees in Apulia Region (Southern Italy). *Phytopathologia Mediterranea* 57(1): 48–72.

Tsing, A. L. 2014. Blasted Landscapes (and the Gentle Arts of Mushroom Picking). In *The Multispecies Salon*, edited by E. Kirksey, pp. 87–109. Duke University Press, Durham, NC.

Tsing, A. L. 2015. *Feral Biologies*. Invited lecture Engagements: The Anniversary of Anthropology in Copenhagen. University of Copenhagen, June 2015.

11 Local cultural values of Persian walnut in Iran

Asaad Habibi, Mohsen Hamedpour-Darabi and Kourosh Vahdati

Introduction

In Iran, walnuts can be found in 28 of 31 provinces. The Persian walnut, as a tree species, has been conserved traditionally for centuries in gardens and orchards throughout numerous provinces of Iran. For this reason, many walnut trees of monumental importance can be found in the country. One of the oldest of these is ca. 1,400 years old and resides in Overgan, a village near ShahreKord city in Chaharmahal and Bakhtiari province (Vahdati 2003; Jafari et al. 2012). Some of these trees produce about 50–100,000 nuts each, while they are believed to be between 300 and 700 years old (Jafari et al. 2012; Vahdati et al. 2014). Once, five limbs of an old tree in Darreka (Rabour, Kerman) came into contact with the ground and grew roots (like layering). In fact, each limb developed roots and branches to resemble a new individual tree. One of these limbs, now a tree, is estimated to be more than 200 years old. A study on the identification of ancient Iranian walnut trees shows that, up until 2016, there were 150 long-lived walnut trees in 27 provinces of Iran, of which 104 trees have been academically recorded and scientifically verified in terms of age (Lotfi et al. 2019).

Walnut trees have good rates of survival and, unlike many horticultural species, they maintain their productivity for centuries. One walnut tree, for instance, is over 400 years old in Yarood village, Qazvin, and produced nearly 65,000 walnuts (fruits) in 2012. This productivity has always ensured the survival of these trees by locals and less interest is given to cutting them down. In the village of Kloce and the surrounding villages in Fereidounshahr, Isfahan, walnut trees are counted as part of the human population of the village. No old tree in this village has a single owner, and every living person can harvest from these trees. With the birth of each child, a walnut is planted and with the death of the person the tree is cut off so that its logs can go to the person's beneficiaries. But the old walnut trees have somehow survived this custom. Since no one owns any of the old trees in the village, no one is allowed to cut any of them down. Hundreds of trees have had the opportunity to continue their life, sometimes reaching absolute maturity (Rezaee et al. 2009; Vahdati et al. 2009a, b; Vahdati and Mohseniazar 2016).

Persian walnut is now seen in vast ranges of geographical distribution, ranging from eastern Europe to the Himalayas and parts of China. Records show that

DOI: 10.4324/9780429320897-13

Figure 11.1 A) old walnut tree in Yarood village, Qazvin province; B) Walnut Game.

the Persian walnut was known in Rome during the reign of Tiberius. In fact, the remains of dried walnuts were discovered in ancient Rome, where walnuts were considered the food of the gods and were named "Juglans regia" in honor of Jupiter. The adjective "English" in English walnut refers to this Iranian dried fruit which the British Marines and merchants took with them on their ships for trade around the world. It is described as the Persian walnut and the black walnut in the eighth edition of the Millers Dictionary, published in 1768 (Tabatabaei et al. 1992: Vahdati 2014).

In Iran, one of the oldest of these trees is 1,400 years old, located in Avorgan village near the city of ShahreKord in Chaharmahal and Bakhtiari province (Vahdati 2003; Ebrahimi and Vahdati 2007). According to an inscription found underground in the vicinity of this tree, the date of its planting and of other nearby walnuts goes back to the time of Yazdgerd III (624–651 AD) (Tabatabaei et al. 1992). Persian walnuts are thought to have been first brought to California by priests on religious missions around 1770. In 1867, Joseph Sexton established the first commercial walnut orchard in California, near Goleta in Santa Barbara County. Evidence also shows that the "Yorka variety" was selected from Persian walnuts which were imported from Iran and Kashmir in 1883 to northern California (Vahdati and Khalighi 2001; Jafari et al. 2012; Avanzato et al. 2014).

Historical documents related to walnuts

Reviewing previous manuscripts in the National Library of Iran (1930) shows that Iran was one of the exporters of walnut kernels to Europe, especially to Britain. These exports included walnut kernels (1942–1942), wood (1934 and 1942) and walnut roots (1936, 1937 and 1938), along with measures that determined the scopes of selling walnut timber and roots. In 1914–1917, a contract was signed for the cutting and sale of Iranian walnut trees in the north of the country between Mohammad Wali Khan, the great lord of Nasr al-Saltanah, and a Russian company through the Mortgage Bank of Iran. In the manuscript numbered document 310/3103, it is stated that in 1931, at the request of the Dikhman Frankfurt Brothers Company in Germany, a permit was issued for the transport of purchased walnut timber from Tabriz, according to the correspondence of the Prime Minister with the Ministry of Welfare and Customs in this regard. In 1953, contracts were signed by the Ministry of Agriculture of Iran and the Iran-Netherlands Export Company to allow the cutting and sale of walnut trees in Astara, Masouleh and Fumanat (document number 320/32010). Also, according to previous manuscripts, in 1970, the export of walnut wood was banned to meet the needs of domestic factories in Iran. In the manuscript, document number 310/48138 (1933), the Cabinet approved the proposal of the General Directorate of Industry and Welfare to issue a license to 380 logs of wood (walnut) from private property and private orchards by Karimi Business House in Khorramshahr to be registered abroad. In another manuscript, it was mentioned that the Council of Ministers, while in agreement with the proposal of the Ministry of Agriculture in 1942, sought to issue a permit for cutting 300 cubic meters of walnut logs for the commercial use of the British Embassy (document number 310/47089). In 1952, according to the manuscript, numbered 4710/47250, the prevention of timber smuggling, the export of oak, walnut and boxwood was registered with the approval of the Council of Ministers in Iran, addressing the export of timber and the fight against timber smuggling.

Social context of walnut trees across Iran

Walnuts can carry with them economic, social and cultural connotations and meanings, although it is essentially a horticultural product. It can be said that walnuts are a component of some rural communities. Walnuts are more than merely an economic issue because of their long-term planting in Iran and the long life of the tree. They can therefore play a role in the socio-economic and cultural structure of the community, especially in rural areas. In addition to its economic role, walnut age, date of planting and the longevity of the tree make walnuts a social, cultural actor. In mountainous areas, walnuts have the same role as dates in tropical areas and olive trees in Mediterranean areas. Walnuts are trees with good longevity and, unlike some orchard species such as peaches, cherries, plums, apricots and pomegranates, which have to be cut down after 20 or 50 years as they lose their productivity through age, walnuts maintain their good productively for several centuries. Walnut longevity is usually longer than human life, which results in a series of beneficiary relationships and productivity systems, sometimes leading to the division of ownership of one tree among several successors of a deceased person. In other cases of strong respect for the tree, the land where the walnut tree is planted remains in the possession of the tree only, and no one else. As long as the walnut tree is alive, the owner of the tree is entitled to the land if he does not own the land, but when the tree dies, the owner of the tree will have no rights in the orchard. If the tree dies and has its roots removed, the layers or root sprouts could make a reappearance of growth and then the ownership of the new growth will return to the previous owner of the original tree. This is extremely important in terms of understanding the concepts of ownership. Olive trees in Mediterranean regions and in parts of Europe (Greece and Italy) and Asia Minor (Turkey) are also known to have a similar pattern of ownership. They have aided in transforming the lives of human populations (Papoli Yazdi and Jalali 2019).

Walnut ownership can be transferred by land purchase or endowment, or as family dowry for girls and boys (mostly for a girl's dowry) at family weddings, thereby becoming the property of the bride most of the time (Avanzato et al. 2014). The difference between a walnut dowry, money and gold is that money or gold can only be claimed by a woman at the time of her divorce, whereas the walnut tree becomes the property of the bride from the wedding onwards, and she has exclusive rights to its annual product. This property can range from one tree to 30 (in Khorasan province). In some areas, at the time of birth, baby girls are given walnut trees as a source of wedding dowry supplies for the future. In most regions of Iran, revenues that come from walnut trees are donated to the religious sites; charities spend the annual income of the walnut harvest on holy sites. For example, there is an old walnut tree in Mahdishahr city, Shahmirzad region, Semnan province that is 250 to 300 years old. Its natural canopy structure has made two open angles with a symmetrical display, because of which the tree is called "Kaman-e Rostam" (Rostam's bow). This tree has been dedicated to Imam Hussein (a sacred person in Islam) since modern records began, and the money which is raised from the sale of its product is used for managing costs of mourning ceremonies for Imam Hussein.

Walnut trees have become a mediator of kinship because of their cultural role and longevity. Individuals from a large family or families that are separated for a variety of reasons are unaware of being part of one or more elderly walnuts, sometimes inherited from several generations ago, and will gather together around the walnut trees for harvest. In fact, walnut trees have been, literally, family trees for many families. In Iran, a walnut tree can be shared among children as de facto inheritance that is more flexible than the laws of the country. Walnut harvest requires resilient and determined manpower because the harvest season is short. Social togetherness is required, and assistance is traditional. One of the most important of these features is assistance and co operation. Landowners or tree owners give walnuts to workers as a reward for their hard work at harvest. For instance, 10% of the walnut yield goes to the person harvesting the crop. The taller the walnut tree, the more fertile and fecund it usually is, and the pay for its harvest is more, which makes the walnut tree of bilateral economic importance for the owner and the workers (Papoli Yazdi and Jalali 2019).

Management and maintenance

In some orchard establishments, walnuts are grown as scattered trees along the borders of orchards. But there are regions where walnut trees comprise the main composition of an orchard. In Tuyserkan, Hamedan, walnut is the main business of the people (Avanzato et al. 2014). Walnut trees can be seen even in the middle of a boulevard and, in this city, there is the "Janbazan" square in which a sculptured vase is full of walnuts. In 2020, a sculpture of a hand, bearing walnuts, was erected in Bavanat city of Fars province to show walnuts as an offering. In Baft, Kerman province, there is a square at the entrance of the city named "Walnut square". There is also a square with the same name in Ziaabad, Qazvin Province, as one of the most important walnut cities in the country (Figure 11.2).

There are many traditions among walnut tree growers in Iran. One of these traditions is to have walnut seedlings called Kalagh-Kaar (planted by crows) which are believed to grow vigorously. These seedlings grow from seeds that are hidden in the soil by crows (for future food) and, somehow, the hidden seeds are later forgotten by the crows. The seeds pass the cold season and germinate in spring (Avanzato et al. 2014). There is a belief that crows select bigger nuts and harvest them from the highest level of the tree canopy where the nuts have been exposed to more light and ventilation during the growing season. In fact, these nuts may not be accessible to humans because climbing the full height of a big tree is almost impossible if traditional methods of harvest are used. So, seedlings that are produced from the seeds of a crow's harvest usually have better vegetative and reproductive growth. There has been a belief that walnut seeds

Figure 11.2 a) Walnut square in Tuyserkan, Hamedan province; b) Walnut boulevard in Bavanat city of Fars province; c) Walnut square in Baft of Kerman province; d) Walnut square in Ziabad of Qazvin province.

would germinate better if the tip of the seed is faced upwards in the soil at sowing time. Another traditional method is that farmers put two nuts in a pot which has a hole in its bottom, place the pot under the soil upside down and let the seeds germinate. It is believed that a competition will take place between the two seeds. The strongest will germinate and use the energy of the other's roots, though it could be a misconception, because of natural grafting which is believed to occur between the roots of the two seeds, thereby making a stronger tree with greater productivity.

Avanzato et al. (2014) reported that growers in the past cultivated multiple nuts so that the young seedlings would screw into each other. It caused the fusion of seedlings and enhanced trunk strength. Moreover, in this case, more wood could be produced with higher quality. Another traditional custom is that some growers scar the trunk by an axe, a spike or the tip of a shovel in early spring to cause sap bleeding (as they call it, "black water"). They believe this makes non-bearing walnut trees pass the juvenility phase into maturity and start bearing. In the past, some growers used this method on trees that produced brown kernels to make the kernels lighter. Some growers spread salt under the canopy of walnut trees to make lighter kernel color. Even if this belief turns out to be a fact, it is not recommended for several reasons, mainly because it increases soil salinity. Some farmers believe that if a walnut tree is planted at night during a full moon, it will grow into a short stature and have lateral bearing, with nuts that fill well. However, if they are planted during the new moon, the trees will grow tall and produce valuable timber (Tabatabaei et al. 1992).

Descriptions of uses and applications

Wood: Walnut lumber has a wide array of applications. Walnut heartwood (stem and root) is well-known for its attractive wood, which is dense, hard, tight-grained and polishes into a very smooth finish. Walnut wood color can range from creamy white in the sapwood to a dark chocolate in the heartwood. For centuries, it has been prized by fine woodworkers for its durability, luster and chatoyancy, and it is utilized for making the body of string instruments (such as guitars and sitars), furniture, carvings, veneers, knobs and handles as well as gunstocks. Walnut wood is used for producing woody doors. For example, Narenjestan-e Ghavam (1850 AD), also known as Ghavam House in Shiraz, Iran, is one of the most remarkable buildings from the Qajar dynasty. All of its doors are made of local walnut trees and are decorated with fish shells from the Persian Gulf. Holy places are also adorned with walnut wood. Examples include the woody pulpit of Sufiyan mosque, Fars province (13th century AD), the building of Mansour shrine in Kashan (16th century AD), Jameh-Atiq mosque in Shiraz (10th century AD) and the door of Syed Sultan Nain Shrine in Nain, Isfahan province (date unknown). Walnut wood is traditionally used in a beautiful art called "Persian Moaragh" which has evolved in Iran from the art of inlay. This method of decorating ornaments with wood goes back to the age of the ancient Persians and Egyptians (Avanzato et al. 2014).

Within the Zagros Mountain range, Kurdistan is known for fine walnut wood craftsmanship, particularly from walnut roots. The wooden board on which Takhte Nard (backgammon) is played can be made of walnut wood. Backgammon and chess are part of Kurdistan's handicrafts and are made of the wood of walnut roots by local artists such as Shabab Kurdistani, who has more than 70 years of experience in walnut root-wood and a worldwide reputation for handicrafts as such, while attracting clients from Eastern Asia, South Africa, Canada and the United States of America. Backgammon boards are pieces of artwork on their own and attract the eyes of each viewer. Walnut root wood is also made into other handicrafts, including ash trays, handbags, trays and boxes.

Shell: Crushed walnut shells are used as charcoal for cooking Kebab and for drilling purposes on oil wells. Two factories have been established in Tuyserkan and Kerman for making active charcoal from walnut shells.

Kernel: According to traditional knowledge in Iran, if too much cheese is eaten alone, it will decrease mental function. However, if the cheese is accompanied by walnut kernels on the breakfast plate, the adverse effects of the cheese would be mitigated by the contents of the kernels. In some Iranian villages, pulverized walnut kernels are used in dough for making bread, especially during Ramadan, the holy month in which Muslims fast. Iranian people believe that walnuts are beneficial for brain function. Historically, they have been suggested to be useful in treating Alzheimer's disease and in helping people with strokes or traumatic damages of the skull. Walnut kernels are widely used in the preparation of many kinds of sweets, ice cream, chocolate, cakes, pancakes and local foods (Avanzato et al. 2014).

Walnut has a long history of remedial and medicinal applications in Persia (Iran), as claimed by many old documents in this regard (Moafegh, 971; Abounasri Heravi 1515; Jahanbani et al. 2016; Jahanbani et al. 2018). The therapeutic properties of each walnut component or product will be discussed in a relevant section.

Other Foods: Khoresh-e Fesenjān, or simply Fesenjan, is an Iranian walnut stew from Gilan province, Iran. In many families, it is cooked for dinner on Shabe Yalda (i.e., the time when families gather together to celebrate the longest night of the year (in January) which is synonymous with the beginning of winter. Fesenjan is flavored with pomegranate paste, pulverized walnut kernels and spices like turmeric, cinnamon, orange peel, cardamom and rosebud. It is traditionally made with poultry (duck or chicken) and is understood to energize the nervous system. Fesenjan can be cooked with meat balls or lamb chunks, but it is very unusual to have fish in it. Depending on the recipe, it can have a sweet or sour taste. Fesenjān is served with Iranian rice (Karizaki 2016). Usually, brown kernels of walnuts are sold at about half price in markets. These kernels are sold namely as "Fesenjān walnut". Fesenjan has a long pedigree. Among the ruins of Persepolis, the famous capital of the Persian Empire, archaeologists found inscribed stone tablets (515 BCE) that list the pantry staples of early Iran, which included walnuts, poultry, pomegranate derivatives and, curiously, the key components of Fesenjān (Oktay and Sadıkoğlu 2018).

Another traditional food that is served in Gilan, a province that neighbors the Caspian Sea, is Kabab Torsh (Sour Kabab). It is made of ground beef meat, usually sirloin or tenderloin, and is marinated in a paste made of pulverized walnuts, pomegranate sauce, chopped parsley, olive oil and crushed garlic. On special occasions, steamed rice is served with meat, lamb, chicken, duck or fish, while stuffed with fried onion, herbs and pulverized walnut.

Shalamin and Ash e Reshteh e Torsh is a famous food which is cooked by people in the mountainous regions of Iran, places like Kurdistan. The ingredients of this dish include walnuts. Aash Sumagh is another traditional dish from Tuyserkan. This food is also tasty and popular. Its ingredients include ash, sumac, chickpeas, beans, beets, milled walnuts and garlic.

A traditional cookie which is served for breakfast in the month of Ramadan in Gilan is the "Reshteh Khoshkaar". It is made of rice, flour, sugar, walnut, cardamom, cinnamon, ginger and spice. It is fried slowly in hot oil. Walnuts are also used for making various kinds of sweets like Gayghanak, a delicious confectionary which consists of flour, eggs, yogurt, walnut and doshab (grapevine syrup). It is served usually at breakfast in winter. Nan-e Berenji, or the Persian rice cookie with poppy seeds, is one of the most traditional cookies in the Persian culture. Nan-e berenji is exquisitely nourishing and exhilarating. Its components include rice, bread, rice flour, water, wheat flour, sugar, cumin and walnut (Vahdati and Hosseini 2006; Karizaki 2016).

Walnut is widely used in the province of Kurdistan to make local cookies. Half a walnut kernel is used for filling the center of some candies like Baslogh (Persian walnut pastry swirl) and Noghl which is a sugar-coated walnut and a famous sweet in West Azarbaijan, Iran. Gaz, a famous souvenir from Isfahan, is made of various ingredients including walnut. Walnut kernels are also used in Halva, Nougat, cakes, cookies, ice creams and watery yogurt with cucumber (a special food for summer) in different parts of Iran (Avanzato et al. 2014). A special dessert that is served in the North of Iran (Gilan) with olives is called Zeitoon Parvarde, made of aromatic herbs, pomegranate, salt and walnut kernels. Ajil is made of walnut kernels that are mixed along with other nuts and dried fruits (called "Ajil-e Moshkel-Gosha" or the Problem-Solving Nuts). This mixture of nuts is usually done on the first night of winter in the Persian calendar, corresponding to the 21st of December, the longest night of the year, which is called the "Yalda" or the "Chelleh" night, when families gather in the house of grandparents to eat watermelon and pomegranate after dinner. This sometimes happens in Nowruz as well (i.e., the beginning of the year in the Persian calendar).

In 2021, two traditional dishes were added to the National Heritage List of Iran. The first was a sweet confectionary, "Halva", in which the confectionary dough is added along with walnuts. The second one was the walnut Kookoo, a falafel in which walnuts are added before cooking.

Fresh kernel: Normally, when walnut trees are harvested toward the end of summer, nuts are usually left in a dry environment to render the kernels dry, hence the usual kernel that is marketed internationally. However, some people in Iran like to eat the kernel fresh, meaning that they break the walnut shell shortly after harvest and eat the fresh, soft kernel. Immature walnuts are pickled in vinegar or saltwater and make a delicious recipe. In summer, immature walnuts are harvested before the hull of the fruit splits open and their full kernels are placed in glass containers with salt water. Every five to ten kernels are sold at a high price in a plastic bag called a Faal (fortune) of walnut. Recently, the nutritional value of fresh walnut kernels has been proven scientifically (Habibi et al. 2017a; Chatrabnous et al. 2018a, b; Sarikhani et al. 2021). Their storage and packaging have been technically addressed while considering scientific, hygienic methods (Habibie et al. 2107b; Chatrabnous et al. 2018a; Habibi et al. 2021), indicating that fresh walnut kernels are highly prized among Iranian people.

Oil: One of the distinctive products of walnuts is the oil that can be taken orally, or it can be used in soup and stew, traditional medicine and in industrial or artistic works. In many villages, walnut oil is used for treating pains of the feet and legs (Papoli Yazdi and Jalali 2019). Walnut oil can be used for lubricating the skin and healing wounds. It is used after the circumcision of boys and in the treatment of tapeworms. A decoction made from the fruit can stimulate the production of thyroid hormones. Using nut decoction on a daily basis can thus be beneficial in treating hypothyroidism (Avanzato et al. 2014). Massagers apply walnut oil on their clients after a massage. It helps relieve tension, pacify nerves and relax the muscles.

Leaves and green husks: The leaves and green husks of walnut trees can be used in both cosmetic and pharmaceutical industries. Several studies have proven the antioxidant effects of walnut leaves and green husks. These byproducts are rich in antioxidant compounds and phenolic acids such as syringic acid, gallic acid, vanillic acid, rosmarinic acid, hesperidin, quercetin and juglone (Chatrabnous et al. 2018a; Habibie et al. 2019). Walnut green husks are reliable color agents, and they are used as dyes in relevant industries. A color in Persian rugs is sometimes made of walnut husks.

Villagers call the walnut leaf "warm", meaning traditionally that it increases the metabolism of human cells. Walnut leaves are used as a traditional medicine for people who have body aches. They sleep on a mat covered with walnut leaves, stretch the body, twist and turn to have the leaves soften their bodies. Due to its tannin content, the decoction of walnut leaves is used in the treatment of diarrhea. A decoction made from dried leaves of this plant can help prevent hair loss. Its leaves are used for treating rheumatic pains, fever, diabetes and skin diseases. Walnut leaves, young fruits and husks are also used in dyeing and cosmetics. Walnut leaves can be used as green manure and, sometimes, as animal feed where dried leaves are mixed with straw (Papoli Yazdi and Jalali 2019).

Canopy: Since walnut trees grow tall and wide, their foliage makes good shade and shelter. In most regions of Iran, single trees are planted for the purpose of shade, and this can be especially beneficial in the summer heat during harvest when workers and landlords need to rest under the shade. Meanwhile, sleeping under a walnut tree at night is almost forbidden in Iranian tradition because it can kill the sleeper, parallel to a saying that "the walnut has taken the men". It has been hypothesized that the big canopy of the tree can create high concentrations of CO_2 at night and thus kill the person sleeping under it (Avanzato et al. 2014).

Landscape: Walnuts are monoecious trees with extensive foliage and massive canopies when they grow over 30 years of age. They develop deep root systems which can help tolerate abiotic stress. Walnuts can be used as ornamental trees in parks and urban green spaces, while having good roles in reducing noise pollution (Rezaee and Vahdati 2008; Rezaee et al. 2008: Vahdati et al. 2018). As previously explained, walnut trees appear frequently in urban landscapes in the main boulevards of Tuyserkan, Hamadan.

With their wide root systems, the trees have key roles in protecting soils against destructive factors such as erosion and human activity. In earliest and ancient times, walnut trees were planted because of their strong and deep root systems. Like willows and mulberry trees, their roots can stabilize soil texture, prevent soil erosion along rivers and mitigate floods (Papoli Yazdi and Jalali 2019). Their roots can reach 5–9 m deep and 12 m wide in the soil, thereby preventing soil erosion. Generally, very old walnut trees have proven that minimal attention and conservation efforts are needed to sustain them when they have established themselves after several years in the soil. Walnuts can be planted across many types of terrain and in the wild, not only to prevent land erosion but also to provide horticultural worth and economic output (Avanzato et al. 2014; Vahdati et al. 2018).

Currency: Walnut trees in Iran can be sold singly in whole or as segments of a living tree. In fact, walnut is sometimes considered as a vegetative livestock in Iran. For example, if a grower is in need of cash, he can sell one or more walnut trees, or even a portion of one tree. This has an important role in the economy of rural families. Walnut is unlike apple or pear trees which cannot be sold individually without selling the land they have roots in (Papoli Yazdi and Jalali 2019).

Proverbs: The long life and high survival of walnut trees partly proves that they are resistant to a variety of adverse conditions over time. Their long life has come to be known as a symbol of patience (Papoli Yazdi and Jalali 2019). "Walnut" is used in many proverbs and metaphors in Iran, for example, "Go play walnuts", which refers to someone who does childish things or should mind his or her own business. "Every circular object is not a walnut" (English equivalent: all that glitters is not gold) is used for comparing similar things or persons with different qualities. When "someone is cracking walnut with his/her tail", the person is very happy or has become successful. "Training an unworthy person is like having a walnut on a dome" means it is almost impossible to train an incompetent person (Avanzato et al. 2014). "Put someone's hand in walnut shell" means making things difficult for someone. "Neither break my head nor put walnut on my skirt" means neither bother me nor do me well. "If the walnut tree is tall, how tall is the melon tree!?" refers to a superficial judgment about something (Avanzato et al. 2014). In some regions, for people who are not related to each other or are very distant from one another, the phrase "someone with someone is a partner in a walnut" is used.

Beliefs: The importance and value of walnuts have made this tree permeate the beliefs of Iranian people and, thus, the tree is made sacred to most areas. For example, goblins take refuge under a walnut tree at night so as to avoid being harmed in the dark. It is customary that people who pass by walnut trees mention the name of God. Hot boiling water should not be poured under walnut trees. If this is done by someone, the walnut tree would curse the person. The walnut tree is sacred and dies in its body and soul if someone cuts it down. Walnuts are known to have masculine features.

Games: Traditional games have used walnuts. One game involves mounting a pair of walnuts on one another and trying to hit them by throwing a walnut from a distance (Figure 11.3).

References

Abounasri Heravi, G. 1515. *Agriculture Guidance* (Ershad Al-Zeraah). The old manuscript. http://dl.nlai.ir/UI/cf0c4112-7703-4523-93f0-6811d1e20591/LRRView.aspx.

Avanzato, D., G. H. McGranahan, K. Vahdati, M. Botu, L. Iannamico, and J. V. Assche. 2014. *Following Walnut Footprints (Juglans regia L.): Cultivation and Culture, Folklore and History, Traditions and Uses*. International Society for Horticultural Science (ISHS), Leuven, Belgium.

Chatrabnous, N., N. Yazdani, V. Tavallali, and K. Vahdati. 2018a. Preserving Quality of Fresh Walnuts Using Plant Extracts. *LWT – Food Science and Technology* 91: 1–7.

Chatrabnous, N., N. Yazdani, and K. Vahdati. 2018b. Determination of Nutritional Value and Oxidative Stability of Fresh Walnut. *Journal of Nuts* 9(1): 11–20.

Ebrahimi, A., and K. Vahdati. 2007. Improved Success of Persian Walnut Grafting under Environmentally Controlled Conditions. *International Journal of Fruit Science* 6(4): 3–12.

Habibi, A., N. Yazdani, M. Koushesh Saba, and K. Vahdati. 2017a. Ascorbic Acid Preserved Phenolic Compounds of Cold Stressed Fresh Walnut Kernels. In I International Conference and X National Horticultural Science Congress of Iran. *Acta Horticulturae* 1315: 665–668.

Habibie, A., N. Yazdani, M. Koushesh Saba, and K. Vahdati. 2017b. Limitation of Access to Oxygen for Inhibition of Browning in Fresh Walnut Kernels. In VIII International Symposium on Walnut, Cashew and Pecan. *Acta Horticulturae* 1318: 137–140.

Habibie, A., N. Yazdani, M. Koushesh Saba, and K. Vahdati. 2019. Ascorbic Acid Incorporated with Walnut Green Husk Extract for Preserving the Postharvest Quality of Cold Storage Fresh Walnut Kernels. *Scientia Horticulturae* 245: 193–199.

Habibi, A., N. Yazdani, N. Chatrabnous, M. Koushesh Saba, and K. Vahdati. 2021. Inhibition of browning via aqueous gel solution of Aloe vera: a new method for preserving fresh fruits as a case study on fresh kernels of Persian walnut. *Journal of Food Science and Technology*. 1–10.

Jafari Sayadi, M. H., K. Vahdati, J. Mozafari, K. Vahdati, M. R. M. Mohajer, and C. A. Leslie. 2012. Natural Hyrcanian Populations of Persian Walnut (Juglans regia L.) in Iran. *Acta Horticulturae* 948: 97–102.

Jahanbani, R., S. M. Ghaffari, M. Salami, K. Vahdati, H. Sepehri, N. Namazi Sarvestani, N. Sheibani, and A. A. Moosavi-Movahedi. 2016. Antioxidant and Anticancer Activities of Walnut (Juglans regia L.) Protein Hydrolysates Using Different Proteases. *Plant Foods and Human Nutrition* 71: 402–409.

Jahanbani, R., S. M. Ghaffari, K. Vahdati, M. Salami, M. R. Khalesi, N. Sheibani, and A. A. Moosavi-Movahedi. 2018. Kinetics Study of Protein Hydrolysis and Inhibition of

Angiotensin Converting Enzyme by Peptides Hydrolysate Extracted from Walnut. *International Journal of Peptide Research and Therapeutics* 24(1): 77–85.

Karizaki, V. M. 2016. Ethnic and Traditional Iranian Rice-based Foods. *Journal of Ethnic Foods* 3(2): 124–134.

Lotfi, N., A. Soleimani, K. Vahdati, and R. Çakmakçı. 2019. Comprehensive Biochemical Insights into the Seed Germination of Walnut under Drought Stress. *Scientia Horticulturae* 250: 329–343.

Moafegh, H. A. M. 971. *Structures of Medication Facts* (Al-abnieh and Haghayegh Al-Advieh). The Old Manuscript.

Oktay, S., and S. Sadıkoğlu. 2018. Gastronomic Cultural Impacts of Russian, Azerbaijani and Iranian Cuisines. *International Journal of Gastronomy and Food Science* 12: 6–13.

Papoli Yazdi, M. H., and Jalali, A. 2019. *Walnut: Symbol of Peasant Patience* (in Persian). Semiannual Journal of Indigenous Knowledge. 9: 1–42.

Rezaee, R., and K. Vahdati. 2008. Introducing a Simple and Efficient Procedure for Topworking of Persian Walnut Trees. *Journal of the American Pomological Scociety* 62(1): 21–26

Rezaee, R., K. Vahdati, W. Grigoorian, and M. Valizadeh. 2008. Walnut Grafting Success and Bleeding Rate as Affected by Different Grafting Methods and Seedling Vigor. *The Journal of Horticultural Science & Biotechnology* 83(1): 94–99.

Rezaee, R., K. Vahdati, and M. Valizadeh. 2009. Variability of Seedling Vigour in Persian Walnut as Influenced by the Vigour and Bearing Habit of the Mother Tree. *The Journal of Horticultural Science & Biotechnology* 84(2): 228–232.

Sarikhani, S., K. Vahdati, and W. Ligterink. 2021. Biochemical Properties of Superior Persian Walnut Genotypes Originated from Southwest of Iran. *International Journal of Horticultural Science and Technology* 8(1), 13–24.

Tabatabaei, M., A. Dehlavi, and A. Ahmadi. 1992. *Juglans, Hickory and Pecan*. Publication of Jahad Daneshgahi, Tehran. (In Persian).

Vahdati, K. 2003. *Nursery Management and Grafting of Walnut*, 2nd ed. Khaniran Publisher, Tehran, Iran. 113 pp. (In Persian).

Vahdati, K. 2014. Traditions and Folks for Walnut Growing around the Silk Road. *Acta Horticulturae* 1032: 19–24.

Vahdati, K., and S. H. Hosseini. 2006. Introducing an Innovative Procedure for Large Commercial Seed Lots Stratification in Persian Walnut. *Acta Horticulturae* 705: 355–357.

Vahdati, K., and A. Khalighi. 2001. Persian Walnut Stooling in Iran. *Acta Horticulturae* 544: 527–530.

Vahdati, K., and M. Mohseniazar. 2016. Early Bearing Genotypes of Walnut: A Suitable Material for Breeding and High Density Orchards. *Acta Horticulturae* 1139: 101–106.

Vahdati, K., R. Rezaee, and M. Mirmasoomi. 2009a. Micropropagation of Some Dwarf and Early Mature Walnut Genotypes. *Biothechnology* 8(1): 171–175.

Vahdati, K., R. Rezaee, and M. Mirmasoomi. 2009b. Rooting and Multiplication Ability of Persian Walnut as Influenced by Mother Stock Vigor and Precocity. *Acta Horticulturae* 839: 223–228.

Vahdati, K., S. Sarikhani Khorami, and M. M. Arab. 2018. Walnut: A Potential Multipurpose Nut Crop for Reclaiming Deteriorated Lands and Environment. *Acta Horticulturae* 1190: 95–100.

Part III

East Asia

12 How Nuaulu sago palms feature in debates around the measurement of plant use and valuation

Roy Ellen

Introduction

Nuaulu are one of many peoples located between the Polynesian outliers to the East and the coastal Indian Ocean islands of Sumatra to the West who rely extensively on the sago palm, *Metroxylon sagu*. Other peoples use sago in various ways without it being culturally central. Drawing on a long-term project on the ethnography of *M. sagu* in eastern Indonesia, this chapter seeks to examine critically the problems of describing its usefulness and value in the context of Nuaulu subsistence, landscape definition and social life beyond the market. Cognizant of the long and chequered history of the literature on measuring value in ethnobotany and economic anthropology, it seeks to unfold connections between different approaches, levels and dimensions of 'plant valuation.'

The 'true' sago palm may be said to be *Metroxylon sagu* Rottboell, first described by G.E. Rumphius as *Sagus genuina* (Rumphius 2011, v1: 293–306: figure 12.1). It is a large palm growing to 25 metres, flowering and fruiting on average after 15 years (though more accurately the range is between 11 and 23 years), but also reproducing by sending out suckers (Flach 1997; Schuiling 2009; Schuiling and Jong 1996). There are other sago palms (Ruddle et al. 1978) which yield edible pith and which are culturally important in different places (e.g., *Eugeissona utilis* Becc. and *Arenga undulatifolia* Becc. among Penan Benalui in northwestern East Kalimantan: Puri 2005: 155–162), but none are as productive as *Metroxylon sagu*.

Having fruited, when sago stem starch is at its maximum, a *Metroxylon sagu* palm will die. Fifteen years is short by comparison with other plants we describe as 'trees' and duration over the long term is a significant feature influencing our relationship with and valuation of trees in general. But whether palms are treated as trees or not, plant species assigned to the family *Arecaceae* score much more highly overall in terms of various measures of utility than most other families, excepting the provision of timber. How then might we begin to make sense of the variety of uses and different valuations placed upon the sago palm and its component parts provided by outside observers (whether scientists, anthropologists, traders or economists) and by the Nuaulu themselves?

DOI: 10.4324/9780429320897-15

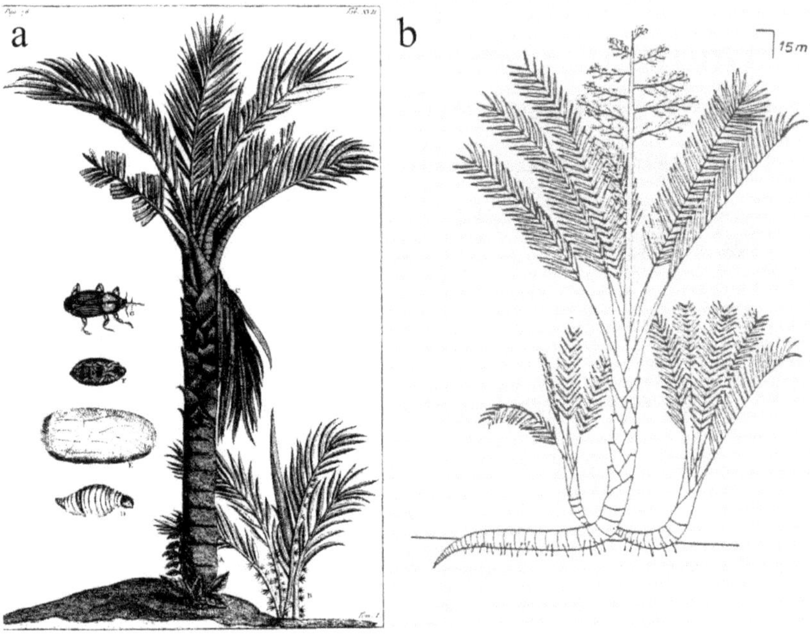

Figure 12.1 The sago palm, *Metroxylon sagu* Rottboel: a) depiction in the *Herbarium Amboinense* (Rumphius 2011, v1: 292, table 17) with image of developmental stages of the palm weevil *Rhynchophorus bilineatus* to the left, and reproduced from the original 1731 edition courtesy of Missouri Botanic Gardens; b) diagrammatic illustration of hapaxanthic and suckering habit, reproduced from Schuiling 2009: 25 after an original by A. de Castro-Dos Santos 1981: 178–179 (Plate 43, figure 12.3).

The sago palm in Nuaulu subsistence and lifeways

In a series of publications (Ellen 1977, 1988, 2004a, 2004b, 2006, 2011) I have documented the role of the sago palm in Nuaulu life. *Metroxylon sagu* remains undeniably the most important Nuaulu plant resource, disease-resistant and present in almost every aspect of their culture, society and ecology. Unlike in many other parts of the Moluccas and southeast Asia more generally, there has been no apparent diminishment in its role as food, and people are apparently unashamed of relying on a product often denigrated as a 'famine food.'

Sago is harvested from palms by felling them about 30 cm from the ground, splitting the trunk, reducing the pith using a bamboo pounder and extracting flour from the pith, operating a device comprising two interconnected troughs (Ellen 2004a). Pith is placed in an upper trough, and water added. The mulch is pressed against a semi-permeable membrane made of coconut leafstalk fibre and the flour passes through in suspension into the lower trough. Once the flour has settled, any surplus water is removed and the flour is packed damp into casks made from sago leaves. The apparatus used has moving parts, often depends on the continuous supply of water from a bamboo conduit system and might accurately be described as a machine. The widespread distribution of similar equipment points to its long history (Ellen 2004b).

The flour is consumed mainly as a gelatinous porridge through the addition of hot water or by baking it in bamboo internodes, sago leaves or ceramic ovens (Ellen and Latinis 2012). It may also be mixed with *Canarium* nut, palm sugar, chili and coconut to make pancakes and sweetmeats (Ellen 2019; Ellen and Goward 1984). The palm is additionally a source of proteinaceous larvae of the black palm weevil, *Rhynchophorus bilineatus*, harvested from the rotting boles and trunks. Of the non-food uses, the most salient is the use of leaves for thatch, the narrow parts of the leaf stalk or petiole-rachis for house walling and other purposes, and the epidermis of wider (lower) parts of the leafstalk as material for making containers (Ellen 2004a).

By and large, placement and arrangement of sago populations in Nuaulu territory should be understood as highly anthropogenic. Most sago palms still grow in their greatest concentration in semi-managed riparian swamp forests. In south Seram these lie, for example, along the banks of the Nua and Ruatan rivers at their confluence, and at the mouths of the larger rivers flowing to the South (such as the Yala) and emptying into the Banda Sea. But palms are also grown in the village area and on garden land. Depending on location, ownership patterns will vary, but even in large areas of swamp forest, where many sago palms are never harvested, these (*hata ipan* or *ipane hatana*) are considered to belong to clans that originated in the area, and since 1990, there have been clashes with incoming transmigrants from nearby islands over access, resulting in deaths (e.g. Ellen 2002). In gardens and groves closer to the villages, ownership may be expressed at a clan, local descent group, house or individual level depending on context, displaying a flexible notion of 'serial' rights. Germplasm in the form of suckers is moved around and selected for, such that 15 or more named folk varieties are recognized based on appearance, flavour, pith processing qualities, growth rate and productivity (Ellen 2006).

Beyond use value in ethnobotanical theory

In ethnobotany and economic botany it has become standard to employ a tight empirical definition of what 'use' and 'value' constitute, with established protocols to assess the number of uses a plant has (e.g., Cook 1995; see also Ellen 2020: 100, 110–111). Such protocols permit enumeration and other kinds of measurement, while meeting the requirements of science and the politically driven accountancy of commercial and state applications. At the same time the process gives a kind of equivalence to all uses without fully evaluating the differential importance of each, inclining to 'flatten-out' the use profiles of particular species. Listing uses has become a bit like the listing of plants themselves, a linguistic habit that makes more sense in a culture based on literacy than it does where orality predominates. So, how we assess the uses and value of Nuaulu sago palms as a natural resource depends on how we define 'use', 'utility' and 'value'.

If we are drawn to extending conventional models to accommodate the importance of valuing nature in all its dimensions, we must apply and evaluate new models in various research contexts. We can begin, perhaps, by distinguishing 'use' from 'value', where the idea of 'utility' as employed in economics (referring to the total satisfaction received from consuming a good or service) acts as a kind of conceptual bridge between the two. Accordingly, Nuaulu seldom think in terms of fixed finite species uses, nor do they go around constantly 'calculating values' of plants in some spurious accounting exercise. Is a mention of the sago palm in a myth or employment of a sago sucker to explicate kinship relations a 'use'? Are uses of all kinds fully commensurable in the usually accepted sense, even though they may be connected to an overall appreciation of the position of a plant species for a particular population? Nuaulu use many resources flexibly and opportunistically, and claim that all plants have uses, and that particular plants can have multiple but often ad hoc and indefinite uses, not all of which may be the best application of their properties. And what is useful often depends on what is available, and 'ranking' of either uses or species is generally not spontaneously evident.

Nowadays, local economies everywhere distinguish market-like transactions from other social domains of exchange and consumption (such as gift-giving) for the same biocultural object, which can move between spheres of meaning and simultaneously exist within several different spheres depending on context. We need to remind ourselves that useful plants are always both biological and cultural artefacts, and by definition are socially embedded. Although broadening the range of uses considered and bringing technical sophistication to how we measure them (e.g., Balick's (1994) 'multiple resource curves'), approaches from quantitative ethnobotany have yet to strictly define or examine the concept of use itself. Indeed, the historic framework of economic botany, and by implication ethnobotany, can only ever provide a limited basis for developing a theory of value. Instead, we might turn to economic anthropology for ideas (Ellen 2017b).

One attempt to integrate different measures of usefulness in order to establish a broader utility, following Turner (1988), has been the index developed by Stoffle et al. (1990) for measuring the ecological and cultural significance of biological resources. This attempt to integrate different measures into a single 'index of cultural significance' combines intensity of use (frequency), exclusivity of use (preference for a purpose) and quality of use (number of uses and parts used). A variation on this theme, particularly in respect to the measurement of resilience, is the number of functional domains in which a plant appears (Ellen 2009). If a plant is a major food, and a source of fibre for manufacturing, and has regularly used medicinal properties, we might predict that it would survive longer as a 'useful plant' than if it only appeared in one of these domains. Another attempt to bring the totality of social and ecological uses together – to create a measure of 'utility' – has been the notion of cultural (biocultural or ethnobiological) keystone species (e.g. Garibaldi and Turner 2004). This extension of the ecological keystone species idea is seen to apply where a species is a key to a local culture, economy and ecological system in a way that maintains its parameters at a useful level, which may or may not be appreciated by any one particular local population.

While all these indices constitute measures of direct and indirect etic utility, they do not always take into account other significant local (non-scientific and emic) perceptions of value. An individual population can produce, exchange and consume an enormous amount of a species by weight or volume but not value it highly, while species may be disproportionately valued in comparison to what is actually produced, exchanged or consumed. This is because there are several and sometimes divergent senses of value in operation. If, as Graeber (2001: 1) suggests, the term 'value' is used in three ways – in reference to what is good (what we usually mean by 'social values'), in the economic sense and in the linguistic sense of 'meaningful difference', we might wonder where in this scheme we place values in the mathematical sense (e.g., what is the value of x?), or how we resolve it with the 19 senses of value offered by Roget (1987: 1125). Indeed, we might ask whether these are really all aspects of the same unified idea? Graeber does not touch on the valuation of plants at all, which is a pity since he would probably have had something interesting to say about it beyond the fact that most approaches are reductionist, naive and essentially neo-liberal.

In developing an explicit theory of 'folk biological value' by drawing on the literature of environmental ethics, the anthropology of value and current debates about ontology, Wall et al. (2019) expand on the simple but crucial question for those who persist with ever more elaborate indirect methodologies: why not ask people themselves?

A heterogenous mosaic of interlocking biocultural dependencies

Metroxylon sagu is the means through which an entire landscape has been established. In particular, it is only through continuous human intervention that a particular pattern of clonal diversity and distribution of palms is maintained. If palms were not cut before fruiting and propagated through removing suckers, much of the patterned diversity would simply be re-absorbed into the genome. While Nuaulu constantly attempt to clonally reproduce a particular kind of sago, its existence in particular places can be subverted by the intrusion of pollination and sexual reproduction. The separateness of folk varieties can therefore only be ensured where they can be tended more intensively than in more extensive and remoter sago swamps. Such places are on the village fringe or in gardens, although these are not ideal places in which to grow sago palms and where yields are almost always low and the condition of the palms less healthy.

This heterogeneity under curation holds more intentional variation still, some of it quite specialized. Different parts of the sago palm used for different purposes translate into different kinds of value: stem starch for food (mainly consumed as porridge and biscuits), fronds for thatch, leafstalks for walling, epidermis of the leafstalk base for containers and trays (*topae, tanopae, tasi*: figure 12.2), spines in sorcery, weevil larvae as a source of protein, larger sections of stipe for fencing and mud-free pathways. Moreover, there is a kind of involution whereby to maintain the sago economy parts of the palm itself are necessary to underpin the production process: in making rafts and bivouacs for use in sago swamp, in the manufacture of troughs used for the processing apparatus (*aha*) and the leaves used for making the casks (*ukura*).

True, it is possible to select a number of quantifiable measures of the usefulness of Nuaulu sago palms. For example, there is the volume of starch produced and consumed per head, the number of calories released per person in order to process sago (63% of total energy intake) and the amount of time spent in extraction (27% of time spent on subsistence activities) (for 1970–71 see Ellen 2011: 50). Then there are the number of named varieties (15), food uses (6), manufacturing uses (40+) and medicinal uses (3+) (Ellen 2006, 2020; Ellen and Goward 1984). However, it is not always meaningful to reduce measures of usefulness to numbers. We can quantify production of some uses but not others, while where there are sufficient numbers of uses to justify statistical treatment, what constitutes a 'use' is not always clear. In considering the pivotal role of sago as a keystone species in the landscape, services such as maintenance of the swamp ecosystem, its role as a carbon sink and in flood control and support for other kinds of biodiversity are evidently important, but calculable only with difficulty. When we reduce uses to numbers, we need to be clear about what units we are counting.

The main action through which the value of sago is established within the Nuaulu population is through consumption within a household, and thereafter the making of gifts of sago flour on marriage and the giving of different kinds of processed sago at feasts accompanying all major rituals. Added value is therefore established through delayed exchange, but not market exchange, and not involving cash value. On such occasions, sago passes between clans connected through marriage and in this context becomes part of an elaborate

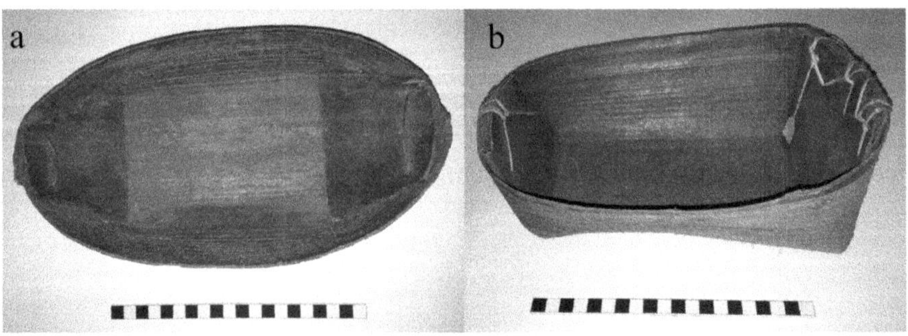

Figure 12.2 Sago leafstalk epidermis used to make containers, Rouhua (*topae, tanopae*): UKC 1970.255 and 1971.528.

set of transactions. Less visible forms of delayed exchange between and within clans are constituted through the labour involved in making sago leaf thatch and walling when building and repairing sacred houses.

Sago is occasionally sold to non-Nuaulu but this is not a significant source of income, and neither do Nuaulu nowadays barter sago products with outsiders. Moreover, the market value of sago casks is now seldom recorded in subdistrict statistical digests of economic data, and the figure for 1982–83 (*Laporan Tahunan Amahai* 1983: 1 kg = 100 Rp = 0.13 USD 1983) is a rare appearance in official sources. In southeast Seram, however, during the 1980s, fish-for-sago exchange between the islands and the mainland was still a major means of placing an economic value on sago. One 'tumang' (cask) of sago was generally the fixed equivalent of two 'ekor' (tails) of dried 'julung' fish (*Hemiramphus marginatus*) in Kwaos, or 100 small 'lemping' (sago biscuits) for one 'waya' (wooden frame) of 'julung' (20 fish) (Ellen 2003: 216–217). Also bundles of sago leaves are bartered for salted dried fish, but in this case are more subject to supply and demand factors.

Value, agency and structure across time and spatial scales

Value is always expressed, experienced and measured locally at one point in time, but the actions that contribute to it derive from a wider spatio-temporal context. If valuations are achieved through a process of exchange, then what emerges at any one moment is a set of values. Like Giddens's (1986) model of recursive structuration, while valuation is always action in progress (agency), it equally always exists in the context of a set of valuations operating at any one time that form the basis for comparison (structure). For the period 1970–2015, when my own fieldwork was conducted, there has been virtually no change in levels of consumption or patterns of use of sago as food, though possibly less production of thatch and leafstalks for building as these were replaced by cement and zinc roofs. However, there has been no decline in the demand for sago-based building materials used for ritual structures, as the accompanying traditional practices remain robust while the population has increased threefold (Ellen 2018).

Just as any assessment of the value of a resource must take into account its durability and changes in its usefulness over time, we cannot fully understand the usefulness or value of a resource to a local population without also looking at its geo-spatial position and how value shifts as we move along chains of exchange. Wall et al. (2019) have sought to develop the idea that valuation shifts depending on the point along the chain where measurement is made, a shift between the valuations of collectors, intermediaries and consumers. Relying in part on value chain studies such as that of Tsing (2015) on the Matsutake mushroom, it is clear that we find different kinds of knowledge (and therefore different kinds of value) at different parts of the supply chain. The knowledge of sea-cucumber collectors in the Aru islands is very different from that of the Chinese traders who purchase them and translates into a very different kind of value (Osseweijer 2000), while the knowledge of medicinal plant collectors from different altitudes and ecosystems in Andean Bolivia is different again from those who sell medicinal products in urban markers, who have knowledge of plant healing properties but not where to find and how to harvest them (Wilkin 2014). Similar considerations apply in the case of sago in the Moluccas.

Sago has been part of a region-wide economy in the Moluccas for some thousands of years. Everywhere, there is local trade to even-out shortfalls in demand and part of its price and value accrues from these local exchanges. However, as we have seen, it has also long been part of a more extensive inter-island trade of the 'fish for sago' kind that is found widely in the Moluccas and Melanesia. On Seram, though only reliably grown in swampy areas, it historically supplied many islands where sago was not grown, but where populations were nevertheless dependent on the products of sago palm for food, house building and manufacturing. Where this interfaced with long-distance trade it was modified to subsidize local economies increasingly growing clove and nutmeg for export to first Asia and then Europe (Ellen 1979). Thus, part of the value of local sago palms is their potential for underpinning the viability or reproducibility of wider systems, which is why I have elsewhere used the notion of 'landscape of exchange' (Ellen 2019) to emphasize how their maintenance depends not only on what happens locally but also what happens in remoter parts of an economic and ecological system. The demand for sago from other islands maintained the ecosystem in the areas of sago production, while there is a demonstrable connection between realms of exchange and movement between sub-economies.

Exchange and the limitations of measuring utility

The Nuaulu word *hunoni* (*huniononi*) covers various meanings connected with the idea of exchange and payment. It can refer to transactions as diverse as those that take place in markets, in the transfer of bridewealth, or adoption of a child in compensation for a homicide. In understanding this semantic range we need to remember that Nuaulu have been part of an international trading system for centuries. In their exchanges, though, Nuaulu make a clear difference between goods obtained through trade (such as gongs and plates that have entered the sacred realm, or fish that have not), and objects that are created outside the market sphere and within a clan, which have value but are never exchanged. Trade objects – such as Chinese porcelain – although they may accrue additional value by virtue of becoming sacred are part of a regime of value where the market has an impact; other Nuaulu heirlooms have no value outside the context of a particular clan or house, such as bark cloth rolls, despite being immensely sacred and full of value by virtue of their association with the ancestors.[1] In this respect, the value Nuaulu attribute to sago is more like the value they attribute to bark cloth rolls. Nuaulu are prepared to kill and take heads to protect patrimonial sago palms (Ellen 2002), an intensity of response not associated with other commercially more important crops, such as durian. While all crop theft is treated seriously, sago theft is always the most serious and surrounded with a rack of supernatural defence mechanisms (Figure 12.3).

Many Nuaulu valuations of sago cannot be understood in terms of what 'others are willing to give up in order to get' (Graeber 2001: 1). The sago palm brings with it a whole set of value implications: its products circulate and underpin social relations in ways and in quantities that other plants do not, for example in inter-clan marriage alliance and in traditional feasting, where it is an essential ingredient accompanying life cycle ceremonies for humans and houses, both as porridge and as biscuits. Nuaulu claim that no meal is complete without sago, that the definition of hunger is the absence of sago (even when other starch sources, such as cassava, are plentiful). When people die, it is a package of sago flour that is placed on top of the corpse to sustain it on its journey to the land of the dead. Unlike many other plant species, no sago folk variety is tabooed or is totemic, all other plants (and many animal food

Figure 12.3 *Wate* (scare charm) erected by Heunaka Sounaue-ainakahata after a suspected theft of sago, Upa valley garden area, near Rouhua, south Seram, February 1996.

Source: (Photograph by author).

species) being divided into those that can be consumed and those that cannot be consumed by various individual people and groups. Beyond the statistical measures of sago utility, the sago palm has a value for Nuaulu that transcends separate or combined uses, though these intangible values clearly depend on the existence of the many tangible uses. The most valued lands that Nuaulu have access to are sago groves held at a clan level, which they have defended in the face of transmigrant insurgency. In a rhetorical defence of outside incursions onto Nuaulu land by lumber companies in 1990, the two theatrical props that Rouhua village headman Komisi employed in an oration to the camera was a clove sapling (emblematic of cash wealth and exchange value) and a sago frond (emblematic of sacred patrimony without direct exchange value) (Ellen 1999).

As a visual entity, the sago palm has no particular aesthetic appeal for Nuaulu, instead the palm is important as a model for understanding the plant world more generally, and in particular for understanding ethnobotanical classification. The physical prominence of sago in daily subsistence privileges it as an interpretative frame of reference and template for other plants, and especially palms. Thus, Nuaulu men are more likely to have learned about plant morphology, growth and ecology from encounters with sago than with any other plant, together with the lexical meronomy that accompanies it. No wonder that it correspondingly features in core origin myths and stories, and in ritual practice, in ways that reflect the cultural centrality of its imagery (e.g., Ellen 2011: 47). According to Nuaulu tradition, it was from the sago palm that Europeans originated.

The consequences of this for plant classification are that its sheer overarching presence sets up a contrast between sago and all other standing palms (not in themselves marked as a clear named grouping), making it the conceptual lens through which other palms are understood. Particular species are perceived as being more or less like sago (Ellen 2020). This conceptual salience is reflected in an internal diversity expressed through 15 or more named folk-varieties, the greatest number for any Nuaulu plant cultigen other than banana which, though common enough, lacks the same kind of cultural currency. Valued sago palms are those more likely to be propagated through removing suckers, tending them at the ends of bamboo water conduits and then replanting elsewhere in a more suitable and convenient location. Nuaulu appreciation of sago diversity as something worth pursuing in itself fits a pattern whereby diversification of objects of all kinds (not just plants) can become part of a cultivated aesthetic indicative of value (Ellen 2017a).

Conclusion

For Graeber (2001: 2) a unified concept of value is arbitrary – 'utterly abstract and vague' – while reconciling the different senses of value is ultimately impossible. Because of the 'value heterogeneity' repeatedly described for different cultural contexts, it is almost as if anthropology here is supplying us with an 'anti-theory' of value in which ethnography constantly subverts practical generalization (Otto and Willerslev 2013). If anything does connect the multiple senses of the word 'value' when we look at biocultural objects such as the sago palm, it is, as Graeber (p. 114) says, those actions and traditions that establish and maintain them, and their comparison with other objects that such actions inevitably imply. Few objects are simply 'one thing or the other', the values attached to any one biocultural object will never all be visible or applicable at the same time, while individual acts of valuation arise from a process in which 'is' and 'ought' are difficult to separate. But as Graeber shows us, we can approach the overall process of valuation for a particular object through several different routes depending on those features we wish to emphasize, just as in ethnography, different interview strategies will capture different aspects of the values we find. One route is via the social biographies of things valued, another through understanding the consequences of the application of human labour and another through patterns of exchange, including the way in which things accrue meaning from transference between persons within whose relationships they are embedded while 'simultaneously detachable' depending on the circumstances (Strathern 1987: 286).

I describe Nuaulu sago palms here as biocultural objects. They are (to follow Ellen 2017b) biological in their phytochemistry but cultural in their use and management, and – following Arjun Appadurai (1986) – with 'social lives' as things. This latter is achieved not only in the ways they are used to think about social relations (as exemplified in Laura Rival's 1998 benchmark collection, *Social Life of Trees*) as well as nature, but also in the way the individual palms or their constituent parts pass through temporal and spatial networks. We can see this simultaneity also in the notion of biocultural keystone species. While certainly a problematic concept both from an ecological and anthropological standpoint, it is one that has proved to be effective in alerting our attention to the comparative significance of species used by particular human populations. Thus, despite their greater monetary value, Nuaulu durian trees cannot be counted as a keystone species, though sago palms indubitably are. By any definition, the Nuaulu sago palm is a biocultural keystone species, and if we are to speak of 'sago societies' as a comparative category which includes peoples as widely dispersed as Siberut islanders (Persoon 1992), Korowai (Stasch 2016), Umeda (Gell 1975) and Nuaulu, then congruity in fulfilling the conditions of this definition is a good starting point. Although there are many populations that depend on the sago palm in various ways, relatively few can be regarded as 'sago societies'. However, this chapter and the earlier studies on which it draws do show us that the idea does not make much sense without seeing *Metroxylon sagu* simultaneously in the context of use, placement (how sago swamp is managed as an anthropic ecosystem), exchange (how germplasm and sago products move around) and how sago palm materials undergo diverse transformations, applications and exchanges not only as food but as construction and manufacturing products as well as through magic. Most societies have multiple spheres of circulation where different measures of value apply. The market may be one of these, but beyond the market are local alternative economies where different principles apply. While it is not unreasonable to treat the sago palm as a complex 'biocultural object', its connected processes of valuation cannot be understood in any unitary way but instead depend on relations established over time and along a socio-geographic arc inscribed by the production, exchange and consumption of its useful parts.

Acknowledgments

The grants underpinning the ethnographic and ethnobotanical research referred to in this chapter are acknowledged fully in the publications referenced. For additional writing support I am grateful to the Leverhulme Trust for its Emeritus Fellowship award EM-2018–057/6. Jeffrey Wall kindly agreed to firmly edit the chapter and reduce its length. His efforts are vindicated by the more focused argument, but he bears no responsibility for any of the shortcomings.

Note

1 Thus, consider such items in a Western auction of ethnographica: while the gongs would fetch considerable sums, rolls of bark cloth would be virtually valueless, despite the fact that both are invested with immensely rich biographical histories as unique and sacred objects.

References

Appadurai, A., ed. 1986. *The Social Life of Things*. Cambridge University Press, Cambridge, UK

Balick, M. J. 1994. Ethnobotany, Drug Development and Biodiversity Conservation – Exploring the Linkages. In *Ethnobotany and the Search for New Drugs*, edited by G. D. Chadwick and J. Marsh, pp. 4–18. Wiley, Chichester.

Cook, F. E. M. 1995. *Economic Botany Data Collection Standard*. Royal Botanic Gardens, Kew.

De Castro Dos Santos, A. 1981. *L'appareil végétatif des monocotylédones: une essai de synth'ese*. Thèse présentée à l'Université des Sciences et Techniques du Languedoc pour obtenir le grade de Docteur de Troisième Cycle Ecologie Génerale et Appliquée, option Ecologie Terrestre. Université des Sciences et Techniques du Languedoc, Académie de Montpelier, Montpelier.

Ellen, R. 1977. The Place of Sago in the Subsistence Economies of Seram. In *The Equatorial Swamp as a Natural Resource*, edited by E. K. Tan. Kemajuan Kanji, Kuala Lumpur.

Ellen, R. 1979. Sago Subsistence and the Trade in Spices: A Provisional Model of Ecological Succession and Imbalance in Moluccan History. In *Social and Ecological Systems*, edited by P. Burnham and R. F. Ellen, pp. 43–74. Academic Press, London.

Ellen, R. 1988. Foraging, Starch Extraction and the Sedentary Lifestyle in the Lowland Rainforest of Central Seram. In *Hunters and Gatherers: History, Evolution and Social Change*, edited by T. Ingold, D. Riches and J. Woodburn, pp. 117–134. Berg, London.

Ellen, R. 1999. Forest Knowledge, Forest Transformation: Political Contingency, Historical Ecology and the Renegotiation of Nature in Central Seram. In *Transforming the Indonesian Uplands: Marginality, Power and Production*, edited by Tania Li, pp. 131–157. Harwood, Amsterdam.

Ellen, R. 2002. Nuaulu Head-taking: Negotiating the Twin Dangers of Presentist and Essentialist Reconstructions. *Social Anthropology* 10: 281–301.

Ellen, R. 2003. *On the Edge of the Banda Zone: Past and Present in the Social Organization of a Moluccan Trading Network*. University of Hawaii Press, Honolulu.

Ellen, R. 2004a. Processing Metroxylon sagu (ARECACEAE) as a Technological Complex: A Case Study from South Central Seram, Indonesia. *Economic Botany* 58: 601–625.

Ellen, R. 2004b. The Distribution of Metroxylon sagu and the Historical Diffusion of a Complex Traditional Technology. In *Smallholders and Stockbreeders: Histories of Food Crop Production and Animal Husbandry in Southeast Asia*, edited by D. Henley and P. Boomgaard, pp. 69–106. KITLV Press, Leiden.

Ellen, R. 2006. Local Knowledge and Management of Sago Palm (*Metroxylon sagu* Rottboell) Diversity in South Central Seram, Maluku, Eastern Indonesia. *Journal of Ethnobiology* 26: 83–123.

Ellen, R. 2009. A Modular Approach to Understanding the Transmission of Technical Knowledge: Nuaulu Basket-making from Seram, Eastern Indonesia. *Journal of Material Culture* 14: 243–277.

Ellen, R. 2011. Sago as a Buffer Against Subsistence Stress and as a Currency of Inter-island Trade Networks in Eastern Indonesia. In *Why Cultivate? Anthropological and Archaeological Approaches to Foraging-Farming Transitions in Southeast Asia*, edited by G. Barker and M. Janowski, pp. 47–60. McDonald Institute Monographs, Cambridge.

Ellen, R. 2017a. Is There a Connection Between Object Diversity and Aesthetic Sensibility?: A Comparison Between Biological Domesticates and Material Culture. *Ethnos* 82: 308–330. https://doi.org/10.1080/00141844.2015.1052085.

Ellen, R. 2017b. Distinguished Economic Botanist Award: Plants as Sociocultural Objects: From Economic Botany to Economic Anthropology. *Economic Botany* 71: 201–208.

Ellen, R. 2018. *Kinship, Population and Social Reproduction in the 'New Indonesia': A Study of Nuaulu Cultural Resilience*. Routledge, Abingdon, UK and New York.

Ellen, R. 2019. Ritual, Landscapes of Exchange, and the Domestication of *Canarium*: A Seram Case Study. *Asian Perspectives* 58: 261–286.

Ellen, R. 2020. *The Nuaulu World of Plants: Ethnobotanical Cognition, Knowledge and Practice Among a People of Seram, Eastern Indonesia*. Sean Kingston for the Royal Anthropological Institute, London.

Ellen, R., and N. J. Goward. 1984. Papeda Dingin, Papeda Dingin. . . . Notes on the Culinary Uses of Palm Sago in the Central Moluccas. *Petits Propos Culinaires* 16: 28–34.

Ellen, R., and K. Latinis. 2012. Ceramic Sago Ovens and the History of Regional Trading Patterns in Eastern Indonesia and the Papuan Coast. *Indonesia and the Malay World* 40: 20–38.

Flach, M. 1997. *Sago Palm Metroxylon sagu Rottb*. International Plant Genetic Resources, Rome.

Garibaldi, A., and N. Turner. 2004. Cultural Keystone Species: Implications for Ecological Conservation and Restoration. *Ecology and Society* 9(3).

Gell, A. 1975. *Metamorphosis of the Cassowaries: Umeda Society, Language and Ritual*. London School of Economics Monograph on Social Anthropology 51. Athlone, London.

Giddens, A. 1986. *The Constitution of Society. Outline of the Theory of Structuration*. Polity, London.

Graeber, D. 2001. *Toward an Anthropological Theory of Value: The False Coin of Our Own Dreams*. Palgrave, New York.

Osseweijer, M. 2000. 'We Wander in Our Ancestors' Yard': Sea Cucumber Gathering in Aru, Eastern Indonesia. In *Indigenous Environmental Knowledge and Its Transformations: Critical Anthropological Perspectives*, edited by R. Ellen, P. Parkes, and A. Bicker, pp. 55–78. Studies in Environmental Anthropology 5. Harwood, Amsterdam.

Otto, T., and R. Willerslev. 2013. Introduction: 'Value as theory': Comparison, Cultural Critique, and Guerilla Ethnographic Theory. *HAU: Journal of Ethnographic Theory* 3: 1–20.

Persoon, G. 1992. From Sago to Rice: Changes in Cultivation in Siberut. In *Bush Base: Forest Farm: Culture, Environment and Development*, edited by E. Croll and D. Parkin, pp. 187–199. Routledge, London.

Puri, R. K. 2005. *Deadly Dances in the Bornean Rainforest: Hunting Knowledge of the Penan Benalui*. Verdandelingen van het Koniklijk Instituut voor Taal-, Land- en Volkenkunde 222. KITLV Press, Leiden.

Rival, L., ed. 1998. *The Social Life of Trees*, Berg, Oxford.

Roget, P. M. 1987 [1852] *Roget's Thesaurus of English Words and Phrases*, edited by B. Kirkpatrick. Guild Publishing, London.

Ruddle, K., D. Johnson, P. Townsend, and J. Rees. 1978. *Palm Sago: A Tropical Starch from Marginal Lands*. University of Hawaii Press for the East-West Center, Honolulu.

Rumphius, G. E. 2011 [1741–50]. *The Ambonese Herbal*, vol. 1, translated by E. M. Beekman. Yale University Press and National Tropical Botanical Garden, New Haven, CT and London.

Schuiling, D. L. 2009. *Growth and Development of True sago palm (Metroxylon sagu Rottbøll) with Special Reference to Accumulation of Starch in the Trunk: A Study on Morphology, Genetic Variation and Ecophysiology, and their Implications for Cultivation*. Unpublished Doctoral Dissertation, Wageningen University.

Schuiling, D. L., and F. S. Jong 1996. *Metroxylon sagu* Rottboell. In *Plant Yielding Non-seed Carbohydrates*, edited by F. Flach and F. Rumawas, pp. 121–126. Plant Resources of South-East Asia 9. Backhuys, Leiden.

Stasch, R. 2016. The Iconicity and Indexicality of 'Life' in Korowai Sago Grub Feasts. In *Des êtres vivants et des artefacts*, edited by P. Pitrou, L. Coupaye and F. Provost. Musée du Quai Branly, Paris. (Les Actes de colloques en ligne du musée du quai Branly), mis en ligne le 20 janvier 2016. http://actesbranly.revues.org/669.

Stoffle, R. W., D. B. Halmo, M. J. Evans, and J. E. Olmsted. 1990. Calculating the Cultural Significance of American Indian Plants: Paiute and Shoshone Ethnobotany at Yucca Mountain, Nevada. *American Anthropologist* 92: 416–432.

Strathern, M. 1987. Conclusion. In *Dealing with Inequality: Analysing Gender Relations in Melanesia and Beyond*, edited by M. Strathern, pp. 278–302. Cambridge University Press, Cambridge.

Tsing, A. L. 2015. *The Mushroom at the End of the World: On the Possibility of Life in Capitalist Ruins*. Princeton University Press, Princeton, NJ.

Turner, N. J. 1988. The Importance of a Rose: Evaluating the Cultural Significance of Plants in Thompson and Lillooet Interior Salish. *American Anthropologist* 90: 272–290.

Wall, J. R., T. Okan, C. Köse, N. Köse, and E. B. Aksoy. 2019. Folk Biological Value and Chestnut Conservation in Turkey. *Economic Botany* 20: 1–16.

Wilkin, P. J. 2014. *Transmission and Commoditisation of Medicinal Plant Knowledge in the Marketplaces of Oruro, Bolivia*. Unpublished Doctoral Dissertation. University of Kent, Canterbury.

13 Revisiting the folk value of Kitayama Sugi under contemporary cultural changes

Andrea Flores Urushima, Benoit Jacquet and Wil de Jong

Introduction

Cryptomeria japonica, or Japanese cedar, is called *sugi* in the Japanese language. Cedar is one of the most valued trees in Japan because it grows straight and is easy to work with, resulting in its widespread use. Cedar has been traditionally produced from the North to South of Japan, but the silviculture techniques used in each region vary, resulting in different tree denominations according to each region. This chapter will focus on the Kitayama cedar, or Kitayama Sugi, the radicans variation of the *Cryptomeria japonica* D. Don (Japanese cedar or cryptomeria), a tall conifer tree that has been produced in the northern mountains of the Yamashiro basin (Figure 13.1), where the city of Kyoto is located. The origins of the working landscapes that developed related to the cultivation of this tree in Kitayama are estimated to have been specially established from around late 17th century, during the period of Sakoku (or the period of Japanese isolationism). The Sakoku autarky regime established the basis for intensifying the use of local human and natural resources, which later launched the foundation for the fast and successful modernization of the country during the Meiji period (Sugihara 2004). Based on a pre-modern taxation system and detailed cadastral and census surveys that assured the control over lands and people, the isolationist strategy spurred a slow growth of the national market, commercialization, and proto-industrialization. This process paralleled with the rise of cultural standards, including refined aesthetic practices, high literacy rates, and the study of mathematics in urban areas. It also allowed for the development of labor and managerial skills in rural areas that resulted in an overall accumulation of human capital. Social stability, self-sufficiency, and high cultural standards assured the development and maintenance of practices and lifestyles which endured long after the end of World War II (WWII). Among those, it is worth mentioning some of the silvicultural techniques and the production of diverse regional specialties, including those related to cedar or *sugi* plantations.

In line with this, some varieties of *sugi* have been grown intensively just north of Kyoto city in a region identified as Kitayama for centuries, but their production intensified after the end of WWII. The tree species has been largely planted in the city suburbs and mountain slopes, as it was used for the construction of

DOI: 10.4324/9780429320897-16

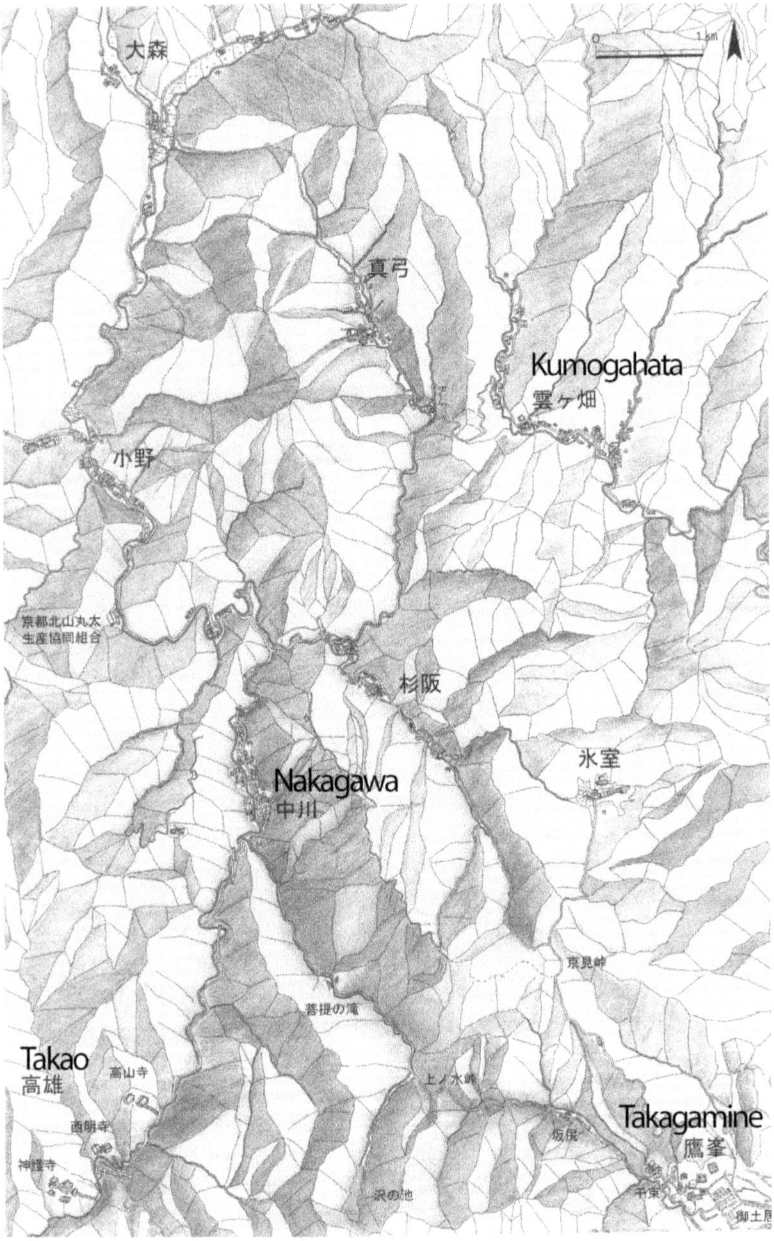

Figure 13.1 Schematic drawing showing the mountains and valleys of the area that concentrates the production of Kitayama Sugi.

Source: Image adapted from (Kyoto City 2019).

Figure 13.2 Landscape of a Kitayama Sugi plantation.

Source: (Photo: Andrea Urushima).

traditional architecture until the end of the 20th century. In Japanese traditional architecture, it is mostly used in what is referred to as Sukiya style architecture, a style that is highly valued throughout Japan, and especially in Kyoto, with its many historical tea pavilions, palaces, and temples. Because of how the tree is used in Sukiya architecture, Kitayama Sugi is grown by applying unique silvicultural methods, including the *daisugi shitate*. Today, the Kitayama Sugi tree, and more specifically, the Kitayama cedar log (Kitayama Maruta), is given the label "controlled designation of origin." It is mostly used in the restoration of tea pavilions and other buildings in the Sukiya style. The enduring appreciation for the tree includes a desire to preserve the aesthetics of rooms, houses, and other buildings built in the Sukiya style, as well as the preservation of related knowledge and practices, along with the social organization formed surrounding the Kitayama Sugi production. This includes, for instance, the preservation of the original mother tree from which the original Kitayama Sugi plantations originated, the preservation of the cultural landscapes in and around the village where the production and distribution have been centered, in addition to the knowledge and practices related to the cultivation and production of logs. In this chapter, we discuss the local, municipal, and federal interest in designating the Kitayama region a conserved cultural landscape. Further, we bring insights from a Kitayama plantation practitioner to consider the associated living culture, which can motivate and disseminate the conservation interests of those who continue to care about Kitayama Sugi, including the "exotic aspects" of its maintenance.

Methods

The authors of this chapter have researched the thematic frameworks treated in this chapter for more than ten years, and have visited Nakagawa on several occasions. One relevant thematic framework in this chapter deals with cultural landscapes and related efforts to preserve the practices and culture associated with this tree species in this specific area.

Research about the impacts of urbanization on rural and traditional ways of living in Japan demonstrates the greatest conceptual changes in terms of land use and landscape management that occurred during the economic growth period (Flores Urushima 2014a, 2015). The 1960s represented a moment of reflection about rural decay during a critical period of urban expansion in the history of Japan. One relevant theme of the period concerned the appreciation of traditional and natural landscapes outside metropolitan areas, especially an emerging appreciation of these landscapes for their aesthetic or environmental aspects, rather than for their productive value (Flores Urushima 2012). Historic landscapes, exquisite natural scenic sites, beautiful landscapes, and the promotion of historical town development have gradually gained attention in preservation policies worldwide and in Japan. Still, productive landscapes became a main target of preservation from as late as the 2000s with the appearance of the laws for the preservation of cultural landscapes (Flores Urushima 2014b). Following this trend, the analysis of the long-term formation of cultural environments after the interaction of humans with local natural environments slowly gained centrality within the discussions about sustainability. The topic has been discussed in the seminar series about cultural environments held annually at the Kyoto University Center for Integrated Area Studies (CIAS) and the Center for Southeast Asian Studies between 2015–2020 (see for example Flores Urushima 2019a). In Japan, one instrumental concept largely used to analyze the process of the formation of cultural environments is *Fūdo*, also translated as "milieu". This word refers to the dynamic and reciprocal interactions that occur between human activities and natural features of places in the process of shaping particular landscapes and ways of living. However, since the postwar period, mechanisms of technical and cultural innovation, as triggered in processes of urbanization and industrialization, have gradually erased *Fūdo*, or the specific traits of cultures and their related landscapes' features (Flores Urushima 2019b). Kitayama Sugi and the lifestyle that developed in villages engaged in its production are archetypes worth illustrating this discussion, and for the future transformation of similar landscapes. This chapter aims to extend this examination, departing from a systemic approach that considers how the preservation of a landscape is intimately linked to the several instrumental and inherent values nurtured inside and outside a specific location in its long-term process of change. From a systemic approach, a way of living developed surrounding the production of a natural resource, the use of which is connected to a distinctive society within a larger system. The collapse/changes that affected the socio-economic-cultural and political features of the larger system affects the maintenance of the cultural values and the physical features of this particular landscape. Up until recently, the landscape of Kitayama Sugi and its birthplace villages endured through processes of modernization due to a strong cultural resilience and flexibility. This chapter will clarify these processes and evidence the role that culture maintains in the process of preserving landscapes and trees.

Origin story

Villages in Kitayama (the North mountains of the Yamashiro basin) have existed for a long time. From around the 12th century, a subsistence lifestyle based on mountainous activities emerged specifically related to the exploitation of firewood and charcoal. The isolated mountainous environments close to a flourishing capital city nurtured the development of a particular regional culture: the Kitayama Sugi industry developed as one of several emblematic forest cultures of Japan. Local residents and researchers on tea ceremony history suggest that the practice might have started in the 14th century. However, the first records of Kitayama sugi production are dated from the late 17th century. Traditionally in Japan, many rural villages developed unique local silvicultural practices to produce posts, beams, and different kinds of planks destined for the country's housing construction.

In parallel to the appearance of the architectural Sukiya style, Kitayama sugi's history began in Nakagawa village, which is today a neighborhood in the northern district (Kita-ku) of the city of Kyoto. The village of Nakagawa, where the specific technique for the production of Kitayama-maruta originated, was first established as a manor of Ninnaji temple and developed in order to support the temple-related activities and provide mountain products to the emperor and other residents of the capital city. Because of its proximity to the capital, the area surrounding the village slowly developed into an afforestation area, and the main village activities evolved toward a focus on the wood economy. During the late modern period until the 1980s, the rise of prices and interest in Kitayama Sugi led to the formation of a mosaic forested landscape of well-defined areas where cedars are grown in even-aged stands. Until 1929, the production of logs was mainly based on familiar labor with living spaces mixed with the spaces for processing the *sugi* timber. After that period, a process of labor reorganization combined with the emergence of new technologies for forestry and manufacturing stimulated the demand for *sugi* and increased the production volume. Thus, during the first half of the Showa period, the reign of emperor Hirohito 1926–1989, an independent, large-scale, multi-story warehouse was built to shelter the activities of peeling and polishing cedar, drying the logs, and storing processed logs. Subsequently, a process of expansion of the village occurred with the installation of other modern small-scale factories and warehouses in the surroundings of the original village. The steep geography of the mountains surrounding Kyoto, the organization of the spaces for living and for production, as well as the mosaic landscape of the cultivated areas formed a unique type of landscape that differentiates Kitayama from other forest landscapes in Japan.

This particular characteristic of the landscape, of the logs' material features, the way of cultivating and processing these logs by foresters, and the use of them by carpenters formed a unified assembly. The cultural value of trees in the Kitayama area is deeply connected to these several dimensions of the use and production these logs.

Although the area has a long history, and was well-known among a specialized public in Japan, the hamlet of Nakagawa became nationally and internationally famous after having been featured in Kawabata Yasunari's novel about Kyoto, *Koto* (The Old Capital)(Kawabata 1962), which was published in 1962 and adapted three times for film and television (in 1963, 1980, and 2005). The chapter on "the Kitayama cryptomeria" describes the location of this village constructed along a river: "The steep mountains on either side of the Kiyotaki River dropped into the narrow valley. One reason the famous cedar logs were raised here was that the area received ample rain and little sunshine. It was also protected from the wind"(Kawabata 2006). Kawabata's novel remains among the most well-known written records about Kitayama Sugi, in which Kawabata compares the rectitude of this tree's trunk to the rectitude of one's mind and heart as stated by Chieko, the heroine in the novel: "The cedars all stand there straight and beautiful. I wish human heart grew like that." The intensive human labor needed to maintain this densely cultivated forest on steep mountains is what allows the tree trunks to maintain the beauty described in this novel. When Kawabata describes the features of this landscape, he is in fact describing the way of living and the ethics and feelings of the people who maintain that landscape.

The production of Kitayama Sugi has suffered several crises since the end of the 20th century, and Kawabata's novel was already reporting the fading of Kyoto traditional craftmanship, from the urban kimono factories to rural Kitayama Sugi forestry, but the challenge of the 21st century is to find new ways to revitalize this folklore and adapt it to contemporary ethics, feelings, and needs surrounding the valuation of trees and the environment at large.

Key environmental factors

The mountains north of Kyoto are a typical example of many of these various areas that developed in close connection to the wood economy. These areas have similar characteristics to that of territories where soil, climate, topography, sunlight, and specific farming techniques have affected the phenotype of the *sugi* trees. The area includes two major types of landscapes devoted to silviculture. The first type encompasses mountains and valleys, which are the locations where trees destined for construction have been cultivated for probably more than 400 years. Then, slightly further north, are the plateaus where the sheds of cooperatives are located – which are large warehouses to store wood – and the log processing factories to precut the various pieces of wood. For this practice an English term, "precut", has been adopted into Japanese as "*prekatto*". The log processing lines produce cut posts and beams according to their function and position inside the houses where they are being used. The cutting process considers the structure of a house and generates a set of parts that are numbered, such that when they are delivered to the construction site, the house's wooden frame can be erected in only a few days.

Commonly found in the subdivided mountain landscapes of central Japan, cypresses and cedars cover the slopes of ridges that are relatively low, but often quite steep. There the cedar grows quick and straight, even in the valleys, where light is significantly less. After WWII, cedar was widely planted around major cities to provide wood for reconstruction. Post WWII, widescale reforestation resulted in a number of ecosystem disservices (Takahashi et al. 2019), for instance pollinosis (*kafunshō*), allergies caused by the species' pollen, which affects some 30% of the Japanese population, and landslides and driftwood. The domestic wood produced in the north of Kyoto is called Kitayama sugi ("cedar of the North mountain") or Tennen Kitayama sugi ("natural cedar of the North mountain"), which refers to the wood that has dried naturally after being cut in the mountain.

Traditional practice and knowledge

The traditional use of *sugi* and *hinoki*, the cypress version of *sugi*, is governed by a number of customs and rules which can be linked with the mechanical and aesthetical characteristics of the species. Cypress was traditionally reserved for visible and primary structures, but when beams are hidden behind a false ceiling, other less expensive species were used. Cedar has been primarily used in the interiors of buildings. Kitayama Sugi was favored for its pale color and therefore highly appreciated by nobility and the bourgeoisie in non-structural corner posts of ornamental alcoves (*tokonoma*) in tatami reception rooms of residences that were built following the Shoin and Sukiya architectural styles of the 16th century. This architectural style is still highly appreciated in Japan to this day. This explains the high prices of *sugi*, even though its demand has declined sharply in recent decades.

Cryptomeria japonica is used widely in housing construction. The species grows quickly and its trunk grows straight. The wood is moderately soft and easy to process. It is, therefore, especially appropriate for the manufacturing of delicate parts in houses, for instance, traditional sliding doors. The species was widely planted around cities to support the reconstruction of houses after WWII. On mountains, usually, *sugi* is planted on the mountain foot, cypress trees (*hinoki*) are planted in the middle area, and pine trees (*matsu*) are planted on the tops of mountains.

Kitayama logs are processed by peeling the cedar bark, immersing the log in water and washing it, and letting the logs dry. Before shipping them to buyers, logs are polished with sand that is collected in areas nearby. Sand polishing is done by hand only. In the Kitayama region, the branches of *sugi* are always trimmed when a tree is growing to produce trees with knot-free wood. A *sugi* stand typically contains trees that only have a remaining upper crown.

Sugi trees are logged during the summer when the sunrays have a near vertical angle and directly reach the forest floor, and the air temperature is high. The trees are cut during the day of the first full moon of August, if it is not raining on that day. The logging crew climb into the tops of the trees and peel off the bark from the top down, while the trees are still standing. After this operation is completed, trees are cut in a circle. Ropes are used to direct the trees toward the center of the circle, toward each other's direction, until a group of tree crowns meet and form a structure similar to the pole skeleton of the North American teepee. These structures of logged *sugi* trees are left for one week in order to receive plenty of sunlight that will allow the wood to dry naturally. The crowns of the trees continue with their photosynthesis and evapotranspiration, sucking up nourishment and moisture from their sapwood. This contributes to the drying, as during one week, a single tree loses about 40% of its moisture. Trees become much lighter and as a result, can be more easily transported to the village where the logs are hand polished with sand and hot water. The drying process is completed inside storehouses, for three months, starting from October onward. Around December, the wood can be used for construction.

Artificial drying of *sugi* trees is faster, but it relies on electric heating which results in wood of inferior quality. A major difference is that trees that are dried in artificial driers lose their moisture as a result of sweating, during which time most of the essential oils are lost as well. The preservation of natural oils during the drying process increases the wood's durability and preserves the visually appealing characteristics, such as its color and shape, which are appreciated in the decoration of rooms.

Traditional workshops have a characteristic work ethic, in which the training of a master carpenter takes approximately ten years, and all the woods employed, with rare exceptions, are of domestic origin. Work undertaked by these workshops is always very expensive. The contemporary Sukiya master carpenter has an intermediary role, between artist and artisan and the construction sector. Although the aesthetic of the construction works is regulated by well-defined proportions, during the construction of buildings, carpenters adapt to the organization of the constructive pieces, their dimensions and construction details, as well as the choice of tree varieties. In this sense, the carpenter plays the same role as the ancient Sukiya carpenters, from the time since they used to be the sole builders of large residences for the nobility in the past (Jacquet 2015). The master carpenters work in collaboration with several other actors or sub-contractors, necessary to keep this specific market segment alive until now: installers of roofing tiles, tatamis, paper sliding partitions, outer wall finishing, etc.

For instance, one of the most famous of them, Nakamura Sotoji (1906–1997) took part in the construction of buildings, tea houses, and Japanese gardens for wealthy clients around the world, notably for John Lennon and Yoko Ono, as well as for the Rockefeller family in the United States, in collaboration with the architect Yoshimura Junzō (1908–1997). These projects provide the living proof that these artisans capture the imaginary of clients belonging to a rich and often international elite, who hold a certain attachment to the image, even the stereotype, of the excellence of the Japanese tradition. The CEOs of Microsoft and Oracle recently called on Sukiya carpenters, instead of established architects, to build their private residences in Karuizawa and Kyoto. These artisans, whose work encompass cabinet making as well, are also seeking new prospects. For example, the Nakamura Sotoji workshop, one of the principal Japanese manufacturers of wooden furniture, has begun to collaborate with Danish designers under the umbrella of the Kohseki company, to offer a selection of contemporary furniture which is compatible with the aesthetic of traditional architecture.

Similar to carpenters, all range of actors related to the traditional *sugi*-based architecture industry are now trying to apply their accumulated means of work and skills to contemporary demands. From the traditional production of wood or paper to the installation of tatamis or paper sliding partitions, the several steps of production and commercialization of materials, up until the final type of products offered, have been systematically renovated. However, it is clear for all actors that the production of materials is deeply connected to the demand of products, and the challenge posed to the preservation of the Kitayama area is dependent on the preservation of the use of this material.

Contemporary conservation context

Today, Kitayama Sugi continues to be produced in the northern mountains of Kyoto city. The north of Kyoto city is commonly referred to as Kitayama, or Northern Mountains, and the area includes several neighborhoods, that used to be independent villages, where Kitayama sugi plantations concentrate, such as Takagamine, Takao, and Kumogahata. Among them, Nakagawa is still the center of the traditional silviculture of Kitayama sugi trees, their processing into polished logs, and their trade to be used in traditional Japanese architecture, as Kitayama Maruta logs. Nowadays, Kitayama landscapes are incorporated into the modern administrative system as neighborhoods of the Kita district of Kyoto city. When you find yourself in Nakagawa, the city of Kyoto seems a distant place, but the town is only about a half an hour drive (12 miles) from the city's center. The area is recognized as the "birthplace" (*furusato*) of Kitayama Sugi and is part of a larger forested territory.

Japan's 37 million hectare territory is almost 70% covered with woods (Forestry Agency 2014). Forest types differ substantially and include boreal forests in its northeastern Hokkaidō, temperate forests on the main islands of Honshū and Shikoku, and subtropical forests in southwestern Kyūshū. Of the country's near 25 million hectares of forest, 10 million are planted forests (FAO 2020), many of which were densely planted for timber production. The country's planted forests are especially poorly managed, and operations like thinning are not undertaken, resulting in the production of wood that is expensive but of low quality. After the 1960s, since the country allowed the large-scale import of timber from abroad, only 30% of wood and wood products domestically consumed are from national forests, while the rest, that is, tree products used for construction, paper, or heating, are imported from Asia and the Pacific, as well as from the United States, Russia, and Scandinavia (Forestry Agency 2014). The forests in traditional forestry production regions are decaying, since the sector is not economically viable, working conditions are harsh, and mountain valleys are remote. Nowadays, populations of traditional forest villages are aging, and forestry occupations hold little appeal to young people and city migrants who are moving to rural areas.

The rural abandonment is apparent in most rural areas across a country where 93% of the population nows lives in cities. The historical investment in the improvement of urban areas, added to the dependence on cheaper imported natural resources and food, explain this situation. Since 2013, alarming demographic studies, such as the Masuda Report (Flores Urushima 2020), which foresees a net decrease of birthrates in rural areas, have generated debate about the best policies for supporting the revival of villages. The relocation of young people to rural areas is gaining attention, and the national government proposed a new plan that aimed to increase the domestic wood consumption to 40% over 2019, in order to achieve net zero greenhouse gas emissions. The plan aims to increase wood utilization in remodeling and construction of non-residential buildings, but it focuses mainly on supporting the mainstream forest industry, where Kitayama Sugi is excluded, as it maintains a special profile.

Currently, Kitayama Sugi logs are mostly used in traditional architecture, especially for the interior pillars exposed in ornamental alcoves, often found in the construction of tea pavilions, or other types of tatami reception rooms built in Sukiya style architecture (*sukiya-zukuri*). From only one variety of seed, a trunk can have about 50 different kinds of phenotypes, and three-meter-high pillars obtain prices that can be up to 1,000 times more expensive than their imported and industrially processed counterparts (Jacquet et al. 2019).

The large differences in production process and prices between imported timber and domestically grown *sugi* trees suggest a narrow relation between traditional architecture, a strong related social organization surrounding the production and construction techniques, and an ecosystem from where the raw materials originate. The balanced relationship maintains the operation of this particular socio-silvicultural legacy. The two species central to this legacy are the Japanese cypress (*hinoki*) and Japanese cedar (*sugi*), along with some other less emblematic species. They have obtained this important role largely because of their mechanical and aesthetic properties.

Real time considerations for the endurance of Kitayama

Posts of Kitayama Sugi are nowadays primarily employed by carpenters in workshops that still specialize in the Sukiya architectural style, which has been used since the 16th century. Of the many workshops in Kyoto that existed previously, only four remain. The main carpentry contractors that specialize in the Sukiya style are Yamamoto kōgyō in Kinugasa, at the northeast of the city; Nakamura Sotoji kōmuten, near the monastery of Daitokuji; Araki kōmuten, in the Western part of the city, in the western Ukyō Ward; and Yasui Moku kōmuten, in Mukō, a city located in the Kyoto prefecture. They are entrusted with the restoration of architectural heritages – tea houses, Buddhist temples, Shinto shrines, and old palaces – along with occasional new constructions, primarily luxury houses.

Kitayama sugi and the Nakagawa area where the Kitayama Sugi silviculture developed, began to be surveyed in 2004 in order to establish whether this area should be recognized as a cultural landscape, an official term included in the law for the protection of cultural properties. The survey was undertaken by the Ministry of Cultural Affairs as one of nine landscapes under its "Projects for the conservation and practical utilization of a cultural landscape". In 2006, the Kitayama Sugi Forestry Landscape became a "project model" for a later established intensive survey of its landscape, architecture, silviculture, and related cultural and social activities (Kyoto City 2019).

One of the wood processing factories is the Keihoku factory, which, according to its manager, operates for 22 hours per day, approximately 360 days per year. Indicative of the transition of the Kitayama Sugi wood industry is that now, 90% of the timber that the Keikohu mill processes is imported timber. In addition, just next to it, there is a sawmill that used to process primarily cedar (*sugi*), cypress (*hinoki*, *Chamaecyparis obtuse*), and pine (*matsu*, mostly *Pinus thunbergii*), but now has been abandoned. The sad and recurring contrast between these two nearby factories is indicative of the Japan wood industry in general. The foreign timber has replaced domestic species that defined the beauty of buildings from the past, but that are still present in most of the country's traditional architecture.

The preservation of the Kitayama Sugi production has significant meaning and involves several social, cultural, and material aspects related to the use and production of this type of tree. But above all, it is the maintenance of the use and the demand for this type of wood which is indispensable in order to maintain the socio-cultural organization developed in connection to its production. This has been recognized by recent efforts to register the Kitayama forestry practice and the landscape associated to its practice as a cultural landscape. Since 2014, the Kyoto municipality in collaboration with the Nara National Research Institute for Cultural Properties, has engaged in a landscape survey of the area (Kyoto City 2019).

There are multiple reasons why Kitayama was subjected to such a survey. On the one side, *sugi* is playing a major role in the preservation of historical buildings. There exist registers of buildings which incorporate logs of Kitayama cedar in their structure, and several of them are national cultural historical patrimony. In order to assure the maintenance and patrimonial legitimacy of these buildings, the production of Kitayama Sugi logs is necessary. Moreover, the preservation of these buildings supports the maintenance of the profession of traditional carpenters and their carpentry techniques. More than 260 buildings in the Kyoto city and prefecture – including townhouses, temples, and tea pavilions – incorporate Kitayama Sugi logs in their structure. Ultimately, the maintenance of these buildings depends upon the maintenance of the production of these logs. On the other side, the production of this type of log involves the preservation of the DNA of the original mother tree (the Great White Cedar), together with the social and physical organization of the village that produces it, and the knowledge involved in the cultivation and production of these logs. Until the beginning of the 20th century, Kitayama Sugi production was centered on the raising up of trunks from the mother tree, in a process called *Daisugi-jitate* (*Daisugi* means the "foundation cedar" or the "mother tree", and *Jitate* means "tailoring", in reference to the way of controlling the growth of smaller offshoots from one mother tree). At that moment, the *Daisugi* trees intermingled with the Japanese red pines or broad leaf forests, which were used for charcoal production.

Contemporary value of the Kitayama landscape

The best way to witness what Kawabata describes is to visit the place of origin of Kitayama sugi, the village of Nakagawa. In this village, the company Nakagen remains as the last one that still maintains the process of naturally drying logs according to the traditional method, as described by Nakata Osamu, the company's owner. We interviewed Nakata Osamu several times in the past and most recently in August 2020 for this chapter. The following paragraphs report some relevant comments from the interview.

Mr. Nakata represents the fourth generation of a collateral branch (*bunke*) that was formed from a main household family (*honke*) of more than ten generations. His great-grandfather received land and was allowed to exploit the mountain beginning in 1889. At that time, his family engaged in forestry but only a small portion of it was dedicated to the production of *maruta* logs, or finely polished logs. His great-grandfather exploited the lands nearby for maintaining the household. On the occasion of a more specific exploitation of large trees, his ancestor maintained an intermediary role of contracting workers to cut the trees under the orientation of a chief local clerk (*bantō*) called Taniyama Hidejirō. It was his grandfather who put effort into cultivating the Kitayama sugi in order to sell to far away buyers coming from Tokyo, Nagoya, or Osaka. The buyers would arrive on the Road 162 by buses provided by a private company called Smith Bus Company, connecting the village to the main railway in central Kyoto. The buyers arrived with a hammer, known as *kokuin*, which had edges that were carved to form the Sino-Japanese characters (*kanji*) of their names. With this hammer, they could mark the logs they bought, by hitting and marking both sides of a *sebiki* – a fissure deliberately created along the trunk in order to prevent its natural cracking. The particular sound of the hammer hitting the logs could be heard from other rooms in the main house. Whenever the sounds echoed successively ("*kon, kon, kon*") it was possible to know how many of the logs had been sold. At the end of the day, the family would gather together to eat *sukiyaki*, or chicken hotpot. This usually happened during the fall season, which was not too hot or too cold, and facilitated killing the chickens and preparing this delicious dish.

Up until 1955, because of the difficulty in cutting and transporting trees down the mountains, only 20% of the mountain owned by the Nakata family was occupied by Kitayama Sugi. It was after 1955 that its production witnessed an expansion due to the great increase in demand in building urban housing. Up until then, most of the buildings that included an alcove and used *sugi* were tea pavilions, temples, or residences of wealthy people. But during the economic growth period, individual houses or apartments all over Japan began to include a tatami room and an alcove, where the use of Kitayama Sugi was indispensable.

Nowadays, the pace of construction has diminished, and fewer new residential constructions include an ornamental alcove, or when these are built, imported wood is often used. This has progressively led to a decline of this customary industry. Mr. Nakata remembers that at the best moment of the Kyoto Kitayama Maruta Productive Cooperative, it was possible to sell the equivalent of 1.5 billion yen during one year. Now, it is possible to reach a modest amount of approximately 20 million yen per year, which represents a decrease of about 98% in sales. In the past, more than 120 members of the cooperative maintained household income exclusively from forestry. Today, that number has decreased to less than 30. Among them, only eight still produce mainly Kitayama Sugi.

Still, Mr. Nakata asks himself why he has continued this activity with utmost effort and considers his relationship to these mountains. He mentions that during the inspection of his lands, on several occasions, he has been able to grasp the condition of the trees. He knows very well when a tree has a suitable environment in terms of water, sun, and soil quality. He is able to observe how the recent increase in rain poorly affects the trees, and how the typhoons (Figure 13.3) have become more frequent, damaging and toppling trees. This is something that he can feel whenever he walks around the mountains, which he has done since he was a child. In the past, whenever he could see the suffering of a tree, he would make efforts to help it. Now, he has mixed feelings. He wants to take good care of the trees, but he knows that an economic profit return is not to be expected. In the end, forestry is business, and the cost-benefit balance has always existed. Up until now, the act of supporting trees was an activity he would engage in without hesitation. It was always his role to support and protect those trees, and now, he feels puzzled by the actual situation led by economic decay that he would have never imagined before.

When asked about his close relationship with his beloved trees, Mr. Nakata remembers that when he was about 12 years old and attended middle school, whenever there was a snowy Sunday, his father would come to wake him up at 7am. Then, they would go to the mountains and remove snow from trees, to avoid damage. In order to release the weight of the snow from the tree crowns, it was necessary to shake down the snow, an action which required a lot of bodily effort by bumping one's shoulders into the trunks. During the winter, in the mountains of Nakagawa, the sunlight will only hit the forest between 11am and 1:30pm. Within this short period of two hours, the snow will melt slightly which made it the most appropriate moment to

Figure 13.3 Trees in the Keihoku area devastated after the typhoon of 2018. Fieldwork research done in collaboration with Roberta Fontan.

Source: Photo by Andrea Urushima.

entirely remove the snow. Falling snow from the top of a crown could hurt the person shaking the tree. Later, when the trees became logs, Mr. Nakata could identify, through observing the annual tree growth ring, the moments when the snow accumulated on the trees' crowns. He could observe the pain of the tree engraved in the variations of the growth ring for whenever the foresters were unable to remove snow. In several moments during the existence of his company, even when the economic situation was unfavorable, he had this strong urge for taking care of the trees, by observing them and feeling an affection for them.

Mr. Nakata could also remember well how his skills improved over time for peeling the bark of trunks and pruning the lower branches (*edauchi*). At all the phases of touching the trees with his own hands, he could feel how the affection for trees would grow. But the moment he felt most happy was when buyers would come and tell him what a wonderful piece he was able to produce, and how that log was valuable to them. Although it was rare for foresters in Nakagawa to have this kind of direct relationship with buyers, some of his clients would phone back later to thank and praise him again for such a wonderful achievement. In some cases, buyers lacked enough money to pay the market price of the log they chose, but he would still sell to them for a lower price out of the satisfaction of having that piece valued by the customers who invited him to see how his log was being used. And the same feeling would be spread among other workers involved in the process when he told them how these logs were valued. He says that what is important in the Kitayama forestry is the satisfaction of people. There are so many people involved in the process of the production of Kitayama Sugi, each of them with a particular responsibility during the process: planting, pruning, logging, peeling off, carrying logs down the mountain, polishing etc. It is from these varied and specialized knowledges that the logs gained their specific features. In the current age of easy disposal, it is very difficult to express to common people the care of so many people involved in the process of producing Kitayama sugi. What make these trees special is that they are treated as part of a family. During the final phase of polishing the logs with sand from the riverside with bare hands, the sugi growers make the point that the trees are joining another family (*yome ni iku*), which is an expression usually used to say that a woman will become a bride and marry. Mr. Nakata says that the value of these trees is similar to that of his daughters: they were nurtured and taken care of by the family, and they would want them to be loved by those who get them.

Conclusion

The Kitayama case is an example of the endurance of a system that adapted to socio economic, political, and cultural changes since the premodern period. This system demonstrates a capacity to adapt to modernization periods due to its cultural value. It has struggled throughout an extremely challenging process of rural abandonment that is counterbalanced by the emergence of new explorations of cultural landscape designation. For the system, there is endurance of both the instrumental values (economic or utilitarian ones – i.e., related to the construction industry and the efforts to maintain the final product quality by caring the tree) and the inherent values (those related to an environmental ethics of care and appreciation for the intrinsic features of nature, and a culture of affection for the trees).

In the novel written by Kawabata Yasunari, the Kitayama plantations and the Nakagawa village are portrayed as cultural creations. In Japan, since ancient times, people are regarded as trees, and in the novel, there is an attempt to humanize the forest by suggesting that the "spirit" of cedars appears four times in the year, during spring, summer, autumn, and winter. Moreover, the cedars that stand in a straight line are an expression of how much people care for those trees. These forests are a proof of how human beings love that creation, since a Kitayama plantation is not a primeval forest, but rather resembles cut flowers in a vase. In the novel description, when the main character of the novel – who was born in Nakagawa – returns to the village, this encounter between the person and the forest is described as if the life returns to the forest of trees. Her father died pruning cedars and it is her memories and her hidden love for that forest that brought life back to that place (Tsuji 2015).

The remaining community in the area still values these trees for the entire range of human and non-human relationships involved in the process of use and production of Kitayama sugi. In a moment when the instrumental values of these trees is being lost, the culture that sustained the system is also being challenged. For a such long-term enduring system, there is now a moment of reassessment of the fortitude of the cultural values associated to its maintenance.

References

FAO. 2020. *Global Forest Resources Assessment 2020: Main Report*. FAO, Rome, Italy. https://doi.org/10.4060/ca9825en.

Flores Urushima, A. 2012. Réévaluation Des Modes de Vie Rural et Citadin Face à La Degradation de l'environnement: Un Débat National Au Japon, de 1967 à 1972. *Revue des sciences sociales* (Strasbourg) 47: 130–138.

Flores Urushima, A. 2014a. *Forests and Urbanisation in Japan: A Historical Perspective*. Kyoto Center of the French School of Asian Studies (EFEO), Kyoto, Japan.

Flores Urushima, A. 2014b. *Between Nature and Culture: Landscape Preservation, Identity and Universal Value*. Kyoto University Graduate School of Letters, Kyoto, Japan.

Flores Urushima, A. 2015. Territorial Prospective Visions for Japan's High Growth: The Role of Local Urban Development. *Nature and Culture* 10: 12–35.

Flores Urushima, A. 2019a. *Lifetime of Urban, Regional and Natural Systems: Examining Examples from Brazil and Japan*. Kyoto University Center for Information Resources of Area Studie, Kyoto, Japan.

Flores Urushima, A. 2019b. *The "milieu" Confronted to Innovation: The 'Perishability' of the Medial Phenomena in the Discourses about the Postwar Modernization in Japan*. EToS Research Report Series No. 1. Hosei University Research Center for Edo-Tokyo Studies, Tokyo.

Flores Urushima, A. 2020. Dans l'entre-Deux Rural et Urbain : Propositions Pour l'aménagement Du Territoire Japonais de l'après-Guerre. In *Mutations Paysagères de l'espace Habité Au Japon: De La Maison Au Territoire*. Bibliothèque de l'Institut des Hautes Études Japonaises, Paris.

Forestry Agency. 2014. *Annual Report on Forest and Forestry in Japan*. Fiscal Year 2014 (Summary). Ministry of Agriculture, Forestry and Fisheries, Japan.

Jacquet, B. 2015. Japanese or Not? How Do Modern Architects Make Japanese Architecture. *Archiv Weltmuseum Wien* 65: 196–217.

Jacquet, B., T. Matsuzaki, and M. Tardits. 2019. *L'architecte et le charpentier: Une histoire de la construction en bois au Japon*. PPUR Presses Polytechniques, Lausanne, Switzerland.

Kawabata, Y. 1962. *Koto*. Shinchōsha, Tokyo.

Kawabata, Y. 2006. *The Old Capital*. Counterpoint, Berkeley, CA.

Kyoto City. 2019. *Kyōto Nakagawa No Kitayama Ringyō Keikan Chōsa Hōkokusho* [Survey Report of the Forestry Landscape of Kyoto Nakagawa]. Kyoto City, Kyoto.

Sugihara, K. 2004. *The State and the Industrious Revolution in Tokugawa Japan*. Department of Economic History, London School of Economics and Political Science, London.

Takahashi, Y., Y. Suzuki, N. Ohta, S. Kawashima, K. Mogami, Y. Yamashita, A. Kusada, S. Ozu, and S. Kakehata. 2019. Multi-Point Analysis of Airborne Japanese Cedar (Cryptomeria Japonica D. Don) Pollen by Pollen Robo and the Relationship between Pollen Count and the Severity of Symptoms. *Aerobiologia* 35: 635–646. https://doi.org/10.1007/s10453-019-09603-0.

Tsuji, N. 2015. Koto Gen'ei: Kawabata Yasunari No Seimei No Ki [A Note on Yasunari Kawabata: Koto (The Old Capital, Kyoto)]. *Kōbe shinwa joshi daigaku gengo bunka kenkyū* 9: 1–18.

Part IV

Africa

14 Djinn-etics of the argan tree

Navigating hybrid kinship between humans and trees

Romain Simenel, Yildiz Aumeeruddy-Thomas and Ahmed Aarab

Introduction

Since Darwin, reproduction has been perceived as a universal phenomenon for living beings that links them in the same chain of filiation. Implicitly, anthropologists have introduced a rupture between humans and other existing beings concerning the articulation between reproduction, filiation and matrimonial relationship. In creating the studies on kinship specific to humans, anthropology has particularized human reproduction as a social and cultural construct that distinguishes it from the reproduction of other species. This observation raises two questions. Would other animals and plants not be able to impose social or cultural criteria in their choice of partner? This question is not the subject of this article, but it leads us to the second question, the answer to which is the subject of this article, namely, have the kinship systems of human groups been thought out in isolation or have they integrated the observation of the modes of reproduction and the relationships that generate them in other species? Logic would dictate that human kinship systems are the result of a complex set of observations of the relationships associated with reproduction in one's own group, in other known human groups, and in existing collectives of animals and plants. Kinship would finally be the reflection of a vast system of understanding of the reproduction of the beings of this world proper to each collective. But one could say that this point of view is totally non-conventional. The case of the genealogical tree is enough for one to be convinced of this. Although many authors have shown that trees are a genealogical symbol (Russell 1979), almost none have identified the ecological reasons for this. However,

> we are concerned with the ways in which assumptions that lie behind a genealogical model – in particular, ideas concerning sequence, essence, and transmission – structure other modes of practice and knowledge making in domains that lie well beyond what we normally label "kinship".
>
> (Bamford and Leach 2009)

Thus, behind the choice of a species to characterize an aspect of human kinship often lies an ecological specificity noticed by observation and which

DOI: 10.4324/9780429320897-18

makes sense to those who use it to think about their kinship. In a very simple way, it is already possible to notice that the images of trees retained to illustrate genealogical filiation in cultures with monotheistic religions are those of species with a significant root system and/or multiple branches, which is why the yew, among other slender trees, does not appear among them. Similarly, the fruits that symbolize the fertility of women are so because of their characteristics that have attracted attention, like the pomegranate, which comes from the Latin granatus "rich in grains", because of the quantity of seeds it contains. Among the Greeks and Romans, the fig also illustrates the fertility of women, even if the fig sign could have signified more obscene figures. Most likely, the fig's metaphor was partly based on the observation that the Greeks and Romans made on the reproduction of the fig tree by caprification, that is, via the introduction of an insect into the fig's orifice. These few examples illustrate the extent to which each culture relied on the observation of the physiognomy and reproduction of plants to metaphorize the sexuality and kinship of humans.

Beyond these anecdotes, ethnobotanical research has demonstrated the finesse of the terminology of the reproductive diversity of certain trees. This is the case, for example, of the palm tree in the Siwa oasis in Egypt, about which the inhabitants have dozens of terms to name the generic, the categories, and the cultivars (Battesti 2013) or the figs in Morocco (Hmimsa et al. 2012). In the face of such richness, one may question the link between the way of naming the diversity and reproduction of the trees and the way of naming the diversity and reproduction of the human. This link can be seen very clearly in the relationships to trees in Morocco. For example, in southern Morocco, in the lands where the argan tree (*Argania spinosa*) grows, these classifications show these instrumental linkages. This article is based on several years of fieldwork carried out in the region of Essaouira and Sidi Ifni between 2002 and 2018, including numerous interviews with local Berber populations.

In the Arab and Muslim world, and in a very significant way in Morocco, djinn are figures that articulate the reproduction and the filiation of the trees to those of the humans. The djinn are supranatural entities or geniuses, and their existence is currently recognized in the Muslim tradition. The djinn are invisible to humans and can take the appearance of man or animal, among other elements. The world of the djinn is conceptualized as being a duplicate of the human world. In southern Morocco, Berber populations believe that as God's creatures equal to humans, the djinn are intelligent. As individuals, they follow customary rules and build social relationships comparable to those of humans (marriage, conflict, assembly, etc.). As such, djinn are equipped with an anthropomorphic interiority (Simenel 2010). The djinn are thought to reside primarily in non-cultivated areas, such as forests, where they are believed to have their homes and their pets (gazelles, jackals, ibex, etc.). Forests are therefore the domestic space of the djinn instead of the domestic space of humans, which includes houses, gardens, fields, and horticultural areas. Therefore, this contradicts the interpretation commonly accepted by anthropological literature on the Maghreb that considers the world of the djinn to be wild or natural (Aracelli

2017; Dermenghem 1954; Gellner 2003; Westermarck 1926). For Berber populations of southern Morocco, there is no separation between a wild world and a domestic world; instead, there is a difference between a world domesticated by humans and one by the djinn. The djinn have, moreover, a very particular link with the growth of trees since each remarkable tree is often interpreted as the fruit of their intervention. In Morocco, the tree species associated with the djinn are numerous. First of all, there is the "tree of the djinn" or the "tree of Seytane" as it is called in Morocco, the Sodom apple tree (*Calotropis procera*) that is featured in some Arab tales. In Morocco, the oleaster, fig tree, cedar, holm oak, and juniper are the species commonly associated with the djinn, as is the argan tree in the south. Most of these trees associated with the djinn are so because they are ritualized. These rituals organized mainly by women are generally related to the requests for weddings and fertility. It is therefore by following the track of the djinn that this article intends to explore the question of the kinship of trees, taking the argan tree as an example.

The argan tree (*Argania spinosa* L. Sapotaceae) is a tree that is endemic throughout southwestern Morocco, the nuts of which produce an oil that has become very popular over the last decade as a result of national and international development policies and trade (Simenel et al 2009). Argan forests, which cover some 850,000 hectares, represent an ecosystem that harbors a significantly rich Mediterranean flora (Alifriqui 2004; M'hirit et al. 1998) and that is under threat from desertification and pastoral exploitation. This area was classified as a "Biosphere Reserve" by the United Nations Educational, Scientific and Cultural Organization (UNESCO) Man and the Biosphere (MAB) program in 1995, even though some areas of argan forests are still not part of the public ("domaniale") lands of the Moroccan government. The region of the argan tree extends from Oued Tensift in the north, to the Saharan fringe around Oued Noun in the south, thus encompassing a large part of the Sous plain, the southern slopes of the High Atlas, and the northern slopes of the Anti-Atlas. A vestige of the Tertiary era, the argan tree occupied a much larger geographical area than today, as evidenced by the presence of relic groups both in the regions of Rommani and Oujda in the north, and south toward the entrance of the Oued Draa and even further south in the region of Laayoun. The argan tree is a monoecious tree with hermaphrodite flowers, and its fertilization is allogamous (it requires cross-pollination between different individuals). The argan tree has botanical characteristics advantageous for the ecosystem because this tree allows for the development and maintenance of a tree cover in a semi-arid climate. It contributes to the fixation and protection of soils against wind erosion, solar desiccation, and runoff. These advantages are used by the inhabitants of southern Morocco to grow cereals within large areas of savannah-type argan populations in the flat and rich soils, while areas on slopes are kept as forests and pastoral lands. In addition, the various forestry (wood, coal, etc.), fruit (fruit, drupe, fines, etc.), and pastoral (tree leaves, plants under wood, etc.) products are exploited by the local population and constitute an important source of income. The argan tree is thus the fundamental component of the entire management system of the environment and occupies a central position in the sylvo-agro-pastoral system of the inhabitants of the argan regions. From the fruits of this tree, an oil is extracted that is an essential element of the diet of these populations and whose economic value is far from negligible. Finally, the argan tree is assigned an eminently social and symbolic role; it is the argan trees that shade cemeteries, souks, wali graves, weddings, ritual picnics (especially between women), baptism ceremonies, and crimes. The populations of these mountains consider some argan trees as meeting points between men and djinn, these invisible beings whose universe doubles that of humans. The argan tree has an enormous cultural value, making it proof of a close relationship between the human populations and this tree, which is articulated thanks to the reference to the djinn.

We shall argue that the question of how human kinship matches argan regeneration is not raised by any ecological determinism between human culture and the ecosystem. The idea is not to reduce human society to mimicry of nature. On the contrary, the question is how human cultures are inspired by ecosystem reproduction to represent their own kinship. In each ecosystem, there is a multitude of models of reproduction that could motivate metaphor with social relation models. Each culture chooses the model that inspires them most. The ecosystem reproduction is also, in return, impacted by human culture because of selection. Finally, the exact purpose of this article is this dialog between human reproduction and tree reproduction, and how humans understand the link between both.

From djinn trees to genetic "mother" trees

In southern Morocco, local history explains that the djinn were there when the first Muslim saints came to occupy the land. A saint is a man bearing "baraka", or divine blessing, who once died in the area. The Aït Ba'amran, a Tribal group of the Sidi Ifni's area, explains that on their arrival in the country, the saints began to teach the djinn the Koran and convert them to Islam. Through the work of saints, certain pagan djinn (kafir) became Muslims. However, the conversion of djinn is synonymous with their control, because once they become Muslims, they find themselves at the service of the saint, exactly as Koranic students are to their master. Due to his control over the djinn, the saint is then able to control the entire environment. He allocates residences to converted djinn in caves, rocks, holes, cairns, and large argan trees. This is how certain argan trees are identified and selected in a symbolic and ritualistic manner for their size or the abundance of their nut production. These argan trees, which are among the most massive and fertile (some up to 14 meters high and 25 meters wide), are named by the name of the saints who would have planted them or would have simply dozed under their shade. These saints are most often men, but there are also argan trees that are associated with women saints. People, especially women, came to those trees with the purpose of visiting the djinn. The main concern is to get a bit of the baraka of the saint that is possible to get from his agents, the djinn. This baraka is needed for different reasons: it may be to make a wish to find a husband or to get pregnant, or to heal diseases, especially skin diseases such as warts and boils, but also muscle strains. The women most often go there with friends and prepare a meal on the spot at the foot of the tree, couscous in this case. The first cooking of the semolina is offered to the djinn according to their eating habits, as well as the blood of the animal sacrificed for the occasion and whose meat will end up in the couscous. To heal skin problems, the pilgrim sets a nail rubbed on a lesion in the bark of an argan tree or hangs barbed wire in its branches after being rubbed onto the affected part of the body. All of these actions help to relieve infectious diseases. The more a tree is visited, the more it is marked with these ex-votos.

Marked by the curse or baraka of the saints and subject to the guardianship of the djinn, these argan trees are respected and maintained as well as fenced off for protection. Aside from the work of thinning the suckers to ensure full growth of the central trunks, these special argan trees do not have their twigs, branches, wood, or foliage exploited in any way. However, the protection of these argan trees cannot alone explain their surprising scale. Pending the results of future genetic studies, it is possible to hypothesize that the genetic heritage of the argan tree, optimally operating thanks to protective conditions, could explain their biological performance. These argan trees of saints undergo a ritualized method of accessing nuts based on an exclusively female industry. Only women living in territorial communities neighboring the argan tree sanctuary can harvest the nuts during a ritual time in relation to agricultural calendar dates. The protection of saints' trees through the djinn and female rituals leads one to think that those argan trees may have a non-negligible role as a genetic "mother" tree in argan tree regeneration. This logically maintains a meaningful parallel between the role of reproduction in human and tree filiations.

Kinship of argan trees

Geneticists have always had trouble with the argan tree. Several genetic studies have been carried out on the argan tree without providing convincing results, except for the observation of a phenotypic and genetic variability not determined by the geographical criterion (F.Bani-Aameur 1998). The genetic diversity of the argan tree can be seen in a multitude of observable characteristics: the shape and color of the leaves, presence or not of thorns, size and shape of the kernels, and the number of kernels in the same shell. Some argan trees have shells but no kernels. Others, quite rare, are half thorny, half smooth. The argan tree can thus madden those who seek to find a genetic logic of individuals and populations. The local populations have integrated all this diversity in their argan tree terminology, where each particularity is meticulously named. However, so far, genetic research has never taken this knowledge seriously. To this genetic diversity is added a diversity of tree facies and the practices that shape them. From the marabou tree, which can reach up to 15 meters in height and 24 meters in wingspan, to the hedges of argan trees, through trees dwarfed by overgrazing, the diversity of facies of the argan tree, which corresponds to the diversity of material and immaterial uses made by local populations, has built the landscapes of the argan area in southern Morocco.

This proliferation of forms and practices linked to a tree makes sense for local populations thanks to a system of filiation that doubles that of humans. Indeed, selection work is clearly evident through the kinship terminology attributed to argan trees. This terminology in the Tachelhit language is built on the fundamental distinction between two regenerative properties of the argan tree, suckering and the ability to produce vegetative sprouts that can ultimately become trees. It is still difficult to establish statistics on this subject, but it is quite obvious that a significant part of the argan tree population does not come from spontaneous germination but from reproduction by suckering and sprouts. The local populations have obviously noticed this phenomenon and have integrated it into the terminology (Genin and Simenel 2011). Suckers growing from the roots of original trees are called daughters or sons depending on the size (*ilis* or *iwis*), while sprouts that grow at the foot of a tree are named uterine nieces or nephews (*tayaot* or *ayao*). There is a strong recognition in argan tree lineage that uses a common vocabulary with the one of human lineages. This terminology has the function of following the growth of trees by determining their filiation: we can say that a small shrub is the sucker, the "daughter", of a particular tree, or that it is the sprout, the "niece" of an old tree it replaced. The mark of the feminine indicates a small size while that of the masculine indicates a large size. This filiation in no way takes into account the resemblance between the trees; on the contrary, this terminology of kinship makes it possible to memorize the filiations between trees which are shaped differently according to the practices associated with them. This is the case, for example, of the stunted hedges of argan trees that border the boundaries of fields and are often identified as suckers, "sons", of argan trees positioned about ten or twenty meters inside an agricultural plot. To understand to what extent this terminology of the argan tree's kinship plays a crucial role in selection work, we must first show how it integrates the classification of the types of kernels for the purpose of argan oil production.

Locals usually argue the reality of these filiations on the basis of the strong similarity in fruit conformation. Regeneration and exploitation are conceived concomitantly. The criteria for selection are based on a typology of the shape of argan tree shells and kernels. As argan oil preparers, women are specialists at defining this typology. According to local classification, there are long kernels, thin kernels, and others small and thick (*tarzift*, *tazdit*, and *tdnit*). Certain shells are mature (*tekhlouf*) and are easy to break while others are more difficult to break (*tamelhkot*) in order to extract kernels. In this way, trees are identified one by one according to the shape of their kernels. The women classify the argan trees according to the shape of their shells and kernel structures, and also according to the timing of maturation and their capacity to be preserved while waiting to be broken to obtain the kernels. Hard shells have this property. The identification of tree classification by a filiation deeply territorialized allows them to very quickly organize the harvest by type of kernels. According to the experience of women, the sprouts or suckers of a tree that provide long kernels will produce the same type of seed. Regeneration by suckers or sprouts thus ensures better control of argan tree populations than by spontaneous reproduction. This phenomenon is more obvious in the agrarian areas where the ploughing of cereals, through cuts of the argan tree roots, favor the root suckers. This favors making the link between mother trees of a given field and the surrounding sprouts. Spontaneous trees, on the contrary, do not necessarily ensure the repetition of the same phenotypes from generation to generation. The harvest takes place every year in July and August after the fruit has matured. Women first harvest the trees with early fruiting and fruit shells that are easy to break, probably those which are the least resistant, while those which are very hard can remain on the ground before being collected. They do this in order to sell them at a higher price before the kernel value falls at the market once the harvest has finished. They also set aside the hardto-crack nuts, which they put in a bucket of water to soften. If crops are harvested tree by tree, the women will not finish until their handmade wicker baskets are filled with a mixture of kernels of different shapes and sizes, a highly valued diversity that is considered a necessity to give the very best taste for a good argan oil.

So, what does the kinship system that structures the categorization of argan trees by individuals correspond to? A proverb in Tachelhit advises to "prune the sprouts", *ikhsa tbi ayao*, in order to let the tree grow or to allow one of them to become a tree. However, the term *ayao* means also nephew in *tachelhit*, and applies to land inheritance. This adage is translated as "prune the uterine nephews", which signifies disinheriting them for the benefit of one's own children, as they are a potential threat to the preservation of the patrilineal territory (Simenel 2010). The terminology of the kinship of the human is thus based on certain features of the ecology of the argan tree to promote a land tenure system that is adapted to it, namely that of preserving a certain indivision of land in the patrilineage. In all of Morocco, the more the land is preserved united, the more it recovers a holy dimension, that called *baraka*, and the more the genealogy of the patrilineage become notorious. Disinheriting uterine nephews is indeed one of the strategies that may avoid land fragmentation, which does not correspond to the ecology of the argan tree and its development. However, all evidence suggests that in field areas, a greater proportion of the argan population comes from suckers than from basal sprouts, and that's certainly a cultural choice that was expressed in terms of a kinship metaphor between humans and trees. Spontaneous trees growing from a seed are also numerous, but they are considered as daughters or sons of the tree that produced them, exactly as suckers. Most of them first grow as a bush, a system by which one major sprout (the strongest) will grow better and become surrounded by multiple sprouts that protect the latter from the teeth of the goats. They are thus selected and maintained. Therefore, if we follow the terminology of kinship, in the field areas, the proportion of daughters and sons in the population of argan trees is greater than that of uterine nephews or nieces. Translated by the terminology and maintained by the agricultural practices, the development of the argan tree in field areas tends to give greater importance to a filiation by men than to uterine filiation. But the ecology of the tree always tends to return to a sprout dynamic which produces bushes and small trees that are related metaphorically to uterine filiation. It is remarkable to note that this "uterine filiation" of the trees characteristic of the ecology of the argan tree corresponds to that in use in the matrilineal kinship systems of these populations until a few centuries ago. This terminology probably does not date from yesterday and was introduced at a time when matrilinearity was more prevalent in kinship systems than it is today. Given the role of women in the production of oil and in identifying the filiation of the argan trees, there is a reason to think or establish a hypothesis that the land tenure system was originally subject to that of the trees structured around a filiation by women, according to the involvement of their knowledge in the management of argan populations. This is also why men like so much to maintain the analogy between "prune the sprouts" and "prune the uterine nephews" thanks to the famous proverb. By valuing so much the culture of suckers, "daughters or sons", these berber populations design a patrilinear landscape which yet overlaps a matrilineal territory of the types of nuts favored and looked for by women.

The perception of generations of argan trees

So, if individuals are distinguished through terminology, what about populations and generations? The perception of the argan trees in the forest is not made individual-by-individual but by groups. These forests are generally larger undivided spaces compared to fields. Very often, these forests are associated with saints and sometimes with their descendants, the chorfa. The latter are bound to live in indivision and therefore not to divide the land inheritance. They are patrilineal, identifying themselves with the filiation of a holy ancestor who is himself a descendant of the prophet, which gives them a certain baraka, a divine blessing. The original forest cover is still locally attributed to holiness and its growth continues to be interpreted in terms of Baraka. Many oral histories depict natural disasters and fate as the main reasons for forest development. One of them tells how such an argan forest would have appeared after the fall of a saint's argan tree during the earthquake of 1936. As a result of the heavy rains that followed the earthquake, the fruit of the argan tree was carried by wadi flood currents and spread throughout the land. From this holy seed grew a forest, and the argan trees within it are called "nephews" (ayao) of the saint's argan tree. According to the Aït Ba'amran's historical view, the argan tree forest has its origins in an event that was both catastrophic and providential, the earthquake, and in the manifestation of Baraka, meaning the fall of the saint's argan tree and the natural seeding of its fruit. As this story illustrates, the initial growth of the argan forest is always directly or indirectly attributed to divine intervention, and the saint's argan tree is presented as just one part of a larger picture. The origin of a forest of argan trees is thus always to be sought in a holy argan tree, which itself has been identified and preserved as a genetic mother plant thanks in particular to its ritual dimension. However, this particular treatment of the holy tree also applies to its descendants, namely the sacred forests. Just as the holy tree is inhabited and protected by the djinn, so too is the forest made up of its descendants.

The Muslim djinn are true representatives of the guardianship of saints and become the guardians of sacred spaces by requiring any intruder to respect certain environmental requirements decreed by the saint. Abundant stories recall the unfortunate fate of people who have violated the environmental requirements of the "agdal" and eventually suffered the punishment of the djinn. Besides the stories of women transformed into porcupines or ogresses after their intrusion into the forest with the intention of cutting wood, others recall the story of a man who had his eye punctured by the thorn of a tree that he sought to chop down. Another folktale recalls a shepherd's flock struck by a serious illness following his abusive use of an agdal area. In all these cases, djinn are the cause of tribulations by intruders for having violated what is considered their living space, the agdal. Djinn are thus seen as the guardians of sacred spaces; they guarantee their closure to any agro-sylvo-pastoral and hunting abuses caused by man. Environmental protocols are therefore primarily intended to preserve argan trees and forest formations. In order to harvest the fruits of the argan trees in these forests, the women organize rituals similar to those of the holy trees, which consist primarily of sharing a meal with the djinn. These rituals are like a pact of alliance with the djinn, who are in charge of the protection of these forests. When the rituals stop for one reason or another, the djinn finally leave, and the protected character of these forests disappears.

How women use argan tree reproduction to integrate men in the logic of the land

If the reproduction of the argan tree is thought of in analogy with the ideal kinship system for women, it is also the symbol of the relationship between man and woman. In these regions of southern Morocco, it is not uncommon for men to come from elsewhere and marry local women. It is a phenomenon so repeated from generation to generation that men always claim to have an ancestor who comes from elsewhere. In this system, it is the women who embody the home and the link to the land. In this configuration, the ancient system of kinship that evokes the reproduction of the argan tree is also a message that men receive from women. It's a signal to think about how the kinship is related to the logic of the land and its ecosystem and areminder of the roots of kinship, which means that they, women, are the main guarantee of the harmonization between the kinship of human and tree reproduction because they come from the land and men come from outside. It is indeed a semiotic way to integrate those external males to the ecosystem of the land. This metaphorical lecture of the kinship as a dialog between human and tree reproduction is a permanent process of integration of the men into the land.

As a perspective of the pregnancy and resiliency of the argan kinship metaphor, the case of the cooperatives is significant. In the name of sustainable development and gender policies, cooperatives of argan oil production were created more than 20 years ago. In fact, by default in these regions, it is frowned upon for married women to work in cooperatives, and the majority of members of cooperatives are therefore widows, divorced women, or single women without brothers (Simenel et al. 2014). Moreover, these women often have no land and therefore no argan trees, because they are generally disinherited by the males of the lineage in order to preserve the land heritage. However, the cooperatives do not generally have any argan trees. The purchase of fruit is therefore necessary. The regular suppliers of a cooperative can be supplied "locally", on the souks, or with the inhabitants of the region. However, some cooperatives tend to use almonds that do not come from the area where they are located. The argan tree fruit chain is made in such a way that cooperatives in the Essaouira region can obtain almonds from the Tafraoute region and vice versa. The nuts of an argan tree are sometimes sold by wholesalers hundreds of kilometers away from the territory in which they are rooted, and in the meantime, they have been mixed with other nuts, from other trees located in other regions (Simenel et al. 2014). Thus, there is often no direct connection between the women of the cooperatives and the fruit they crush. By employing a large majority of single, widowed, or divorced women, who have little or no rights to the fruit-producing argan trees, women's cooperatives can only disconnect themselves from the local territories and communities in which they are located and which, however, are likely to provide them with the only legitimacy that exists in terms of geographical identification.

This situation leads to a paradox that speaks about volumes about the resilience of the argan kinship metaphor, that of these women without kinship who work to produce argan oil from the nuts of trees that they will never see the physiognomy of. The social rupture thus goes hand in hand with the ecological rupture between a world of production "in development" of argan oil and that of the relationship to the resource (the argan tree). This rupture is reflected in the marginalization of cooperatives in rural areas, first manifested by a reputation tarnished by numerous rumors and gossip revolving mainly around the sexuality of young single workers. The latter were often accused of prostitution by the young men of the region who did not hesitate to tag the walls of the cooperatives with obscene references. The model of the women's cooperative, which is supposed to embody a breath of social and economic emancipation, is ultimately imprisoned by its own approach, in a classic scheme of the relationship to the female gender in the Muslim world, that of opposing the licit to the forbidden. In the Moroccan rural world, the single women's cooperative, the argan oil production space, is then perceived as a halal, licit place, exactly opposite to the resource space, the argan agdal, which is seen as a haram, forbidden, domain of married women. This situation gives us some lessons on the limits of development policies and protection of the tree that do not take into account local representations of the tree.

Conclusion

For the Berber-speaking populations, there is indeed a link between the way in which the trees are affiliated and the way in which they themselves are affiliated. Everything is thus done in such a way that the population of argan trees is managed according to the principle of filiation thought in terms of the kinship of the local populations, from the identification of the mother-foot trees by their ritualization, to the selection of suckers and rejects, and the extension of the holy character to an entire forest. The interweaving of kinship and tree populations suggests in particular the strong bind between the mutually assured continuation of beloved ecologies and culture, with the tree as a widespread defining symbol of this lived meaning. The connection between the filiation of trees and that of humans is, moreover, made possible only through the intermediary of the djinn, essential actors in the ritualization of the ecology of the argan tree. The genetic selection of individuals and populations of argan trees as it was performed by local populations is thus similar to a Djinn-etics of the tree governed by both material and immaterial laws of ecology. We now understand to what extent the facies of the argan tree stands in southern Morocco are the reflection of a hybridization committed between the filiation of humans and trees. It is not surprising, therefore, that cooperatives that produce oil from nuts that have no exclusive affiliation with a particular territory have been able to attract only unrelated women, further marginalizing them when the original intention was to empower them. The sustainable and gender development policies have thus failed to learn the life lesson of these populations, which consists of the fact that when the kinship of the argan trees is broken, that of the women is too.

References

Alifriqui, M. 2004. *L'écosystème de l'arganier*. Programme des Nations Unies pour le développement (PNUD-Maroc), Rabat.

Aracelli, G. V. 2017. L'hospitalité des humains envers les *jnûn*. *Revue d'ethnoécologie* [online]. Supplément 1.

Bamford, S., and J. Leach, eds. 2009. *The Genealogical Model Reconsidered*. Berghahn Books, New York/Oxford.

Bani-Aameur, F. 1998. *L'Arganier et les plantes des zones arides et semi-arides*. Université Ibn Zohr, Agadir.

Battesti, V. 2013. L'agrobiodiversité du dattier (*Phoenix dactylifera* L.) dans l'oasis de Siwa (Égypte). *Revue d'ethnoécologie* [online]. Supplément 4.

Dermenghem, E. 1954. *Le Culte des saints dans l'Islam maghrébin*. Gallimard, Paris.

Gellner, E. 2003. *Les Saints de l'Atlas*. Bouchène, Saint-Denis, France.

Genin, D., and R. Simenel. 2011. Endogenous Berber Management and the Functional Shaping of Rural Forests in Southern Morocco: Implications for Shared Forest Management Options. *Human Ecology* 39(3): 257–269.

Hmimsa, Y., Y. Aumeeruddy-Thomas, and M. Ater. 2012. Vernacular Taxonomy, Classification and Varietal Diversity of Fig (ficus carica L.) among Jbala Cultivators in Northern Morocco. *Human Ecology* 40: 301–313.

M'hirit, O., M. Benzyane, and F. Benchekroune. 1998. *L'arganier. Une espèce fruitière forestière à usages multiples*. Mardaga, Sprimont, Bruxelles.

Russell, C. 1979. The Tree as a Kinship Symbol. *Folklore* 90(2): 217–233.

Simenel, R. 2010. *L'origine est aux frontières. Les Aït Ba'amran, un exil en terre d'arganiers (Sud Maroc)*. Editions CNRS – Maison des sciences de l'homme, coll. Les chemins de l'ethnologie, Paris.

Simenel, R., G. Michon, L. Auclair, Y. Thomas, B. Romagny, and M. Guyon. 2009. L'argan: l'huile qui cache la forêt domestique. *Autrepart* 50: 51–74.

Simenel, R., B. Romagny, and L. Auclair. 2014. Les femmes berbères gardiennes des secrets de l'arganier: le détournement des pratiques locales. In *Genre et savoirs: Pratiques et innovations rurales au Sud* [online], pp. 179–200. IRD Éditions, Marseille.

Westermarck, E. 1926. *Ritual and Belief in Morocco*. Macmillan, Londres.

15 Make way for the 'new king'

Farmers' multifaceted relationships with *Eucalyptus globulus* in northwestern Ethiopia

Morgan L. Ruelle and Zemede Asfaw

Introduction

All over the world, as agriculture expands and forests shrink, farmers plant, protect, and tend to trees (Zomer et al. 2014). An inventory of ecosystem services from trees explains why farming communities value trees in their landscapes (Reed et al. 2017). Provisioning services that are important to farmers include food (fruits, nuts, leaves, oils, and spices); medicines; forage and fodder for domesticated animals; construction materials for barns, fences, and storage facilities; fuel for heating and cooking; as well as gums, resins, and latex used in various ways. Trees also enhance regulating services, including carbon sequestration, flood mitigation, air quality improvement, erosion control, and surface water protection. The supporting services provided by trees include habitat for pollinators and other beneficial organisms; shelter for people and domesticated animals; and soil enrichment to enhance productivity. Lastly, trees provide cultural services, including inspiration for art and innovation; markers of historical events; spaces for social gatherings and practices; sacred sites for religious practice; and meeting places for community governance.

However, trees may also compete with crops and other organisms for space, soil nutrients, water, and sunlight, depending on the species and their spatial arrangement. Farmers manage the abundance and distribution of trees to limit their negative impacts on agricultural productivity (Nath et al. 2016). In many regions, farmers face additional pressure to reduce the number of trees: increasing population densities and higher market demands require producing more food in less space, and trees are often sacrificed to expand croplands. With the exception of regions where trees are major crops (e.g., coffee, rubber, or palm oil production areas), canopy cover on agricultural land is declining across the globe (Zomer et al. 2014).

Given the challenges facing farmers, governments and civil society organizations have been promoting exotic multipurpose trees that can provide numerous ecosystem services in less space and with fewer resources than might be required by the assemblages of indigenous species that provided those services in the past (Franzel et al. 1996; Mekoya et al. 2008). Examples of fast-growing, multipurpose trees promoted in the tropics include *Moringa oleifera*, *Sesbania*

DOI: 10.4324/9780429320897-19

bispinosa, *Grevillea robusta*, *Leucaena leucocephala*, *Casuarina equisefolia*, and various species of *Eucalyptus* (Mekoya et al. 2008). By planting these species, their advocates reason, farmers can meet many of their needs while opening space for crop production and reducing pressure on indigenous trees. Many of these species have been planted by millions of farmers, which may indicate their value. However, given pressures and incentives to plant multipurpose trees, it is important to speak with farmers and gain their perspectives, paying close attention to economic as well as ecological and cultural values.

The notion of 'folk value' introduced in this volume can be applied to understand the relationships between farmers and multipurpose trees. This chapter examines farmers' relations with *Eucalyptus globulus*, a multipurpose species introduced to Ethiopia in the late 19th century. We focus on the Debark District of northwestern Ethiopia, a highland agricultural landscape where most indigenous trees are rare and eucalyptus is the dominant species. Based on interviews and other ethnobotanical research activities with farmers, we compare the values of eucalyptus to those of the once common indigenous tree species of the area. By considering its uses, ecological impacts, and cultural significance, we demonstrate the multifaceted complexity of farmers' relationships with eucalyptus.

Background

Ethiopia is widely recognized as a center of botanical diversity, with an estimated 5,757 plant species (Kelbessa and Demissew 2014). The most recent study found that 5,219 of Ethiopia's plant species are indigenous and 647 of these are endemic (Demissew et al. 2021). An estimated 85% of Ethiopia's population of nearly 115 million in 2020 is engaged in agriculture; most are subsistence-oriented farmers living at elevations higher than 1,500 meters above sea level. Ethiopia's highland communities have been practicing agriculture for thousands of years and have maintained non-domesticated useful trees in their landscape. Contrary to a popular but weakly supported narrative that Ethiopia was deforested in the 19th and 20th centuries, historical analyses suggest that much of the country's natural forests were cleared much earlier (McCann 1997; Wøien 1995). Historical photographs reveal that some parts of the country are more densely forested now than in the early 19th century (Meire et al. 2013; Nyssen et al. 2009). However, afforestation programs in the 20th century included many exotic tree species, including about 55 species of the genus *Eucalyptus* (Dessie and Erkossa 2011).

The introduction of eucalyptus to Ethiopia is attributed to emperor Menelik II, who founded the nation's capital at Addis Ababa in the 1880s. For many centuries, Ethiopian emperors had frequently relocated the seat of government, in part because they had exhausted the timber and fuel resources of surrounding areas. Hoping to establish Addis Ababa as a more permanent capital city, Menelik II turned to his foreign emissaries for advice. One of them – either a French or British advisor – introduced Menelik to eucalyptus, which grew much faster than any indigenous tree species, and could therefore satisfy the city's demand for wood. Within less than a decade, the city had been transformed from a nearly treeless city to a forested one, a 'eucalyptopolis', according to one observer (Pankhurst 1995).

Over the next several decades, eucalyptus spread throughout the country, at first to other urban centers. It wasn't long before detrimental effects were recognized: the tree's rapid growth is achieved by withdrawing large amounts of water, drying up streams, wells, and understory vegetation. In 1913, Ethiopia's nascent Ministry of Agriculture, concerned about its effects on crop production, ordered citizens to uproot two out of every three eucalyptus trees (Pankhurst 1995). Throughout the last century into the current one, Ethiopian biologists, ecologists, foresters, and agricultural scientists have pointed out the negative impacts of eucalyptus on land, water, and biodiversity (Dessie and Erkossa 2011). Some regional governments (e.g., Tigray Regional State) have banned planting of eucalyptus on cultivable land (Jagger and Pender 2003), but the species continues to proliferate throughout the highlands (Mekonnen 2010).

Given that eucalyptus is found throughout the country, it is included in nearly every inventory of plant use (e.g., Chekole et al. 2015), and many previous studies have indicated its contributions to farmers' livelihoods (e.g., Feyisa et al. 2018). However, there has been little attention to the nuances of farmers' relationship with this tree. One explanation is that ethnobotanists tend to focus on indigenous species, and there has been surprisingly little research investigating relations with introduced species. Others have identified the need to engage farmers in discussions about the complex effects of eucalyptus on their lives (Adimassu et al. 2010). The present study applies the concept of folk value to investigate farmers' relationships with eucalyptus, including its use within their communities, its impacts on their agroecosystems, and its sociocultural significance as compared with other tree species.

Methods

Study area

The Debark District is located on the western slopes of the Semien Mountains, at the northern edge of the North Gonder Administrative Zone in the Amhara Region (see Figure 15.1). The center of the District is Debark town, which sits along a major trade route between the cities of Gonder (100 km to the south) and Axum (250 km north and east). Debark town is 2,850 meters above sea level, with temperatures ranging from less than 10°C in January to over 25°C in March. Average annual rainfall is 1,660 mm, following a unimodal distribution; the rainy season extends from April through September and peaks in June and July (WorldClim 2021).

At the time of the last national census, the population of Debark District was 160,130 (Central Statistics Agency 2008). The majority of Debark residents identify as Amhara and speak Amharic as their first language. Most are followers of the Ethiopian Orthodox Tewahedo Church, although there are significant

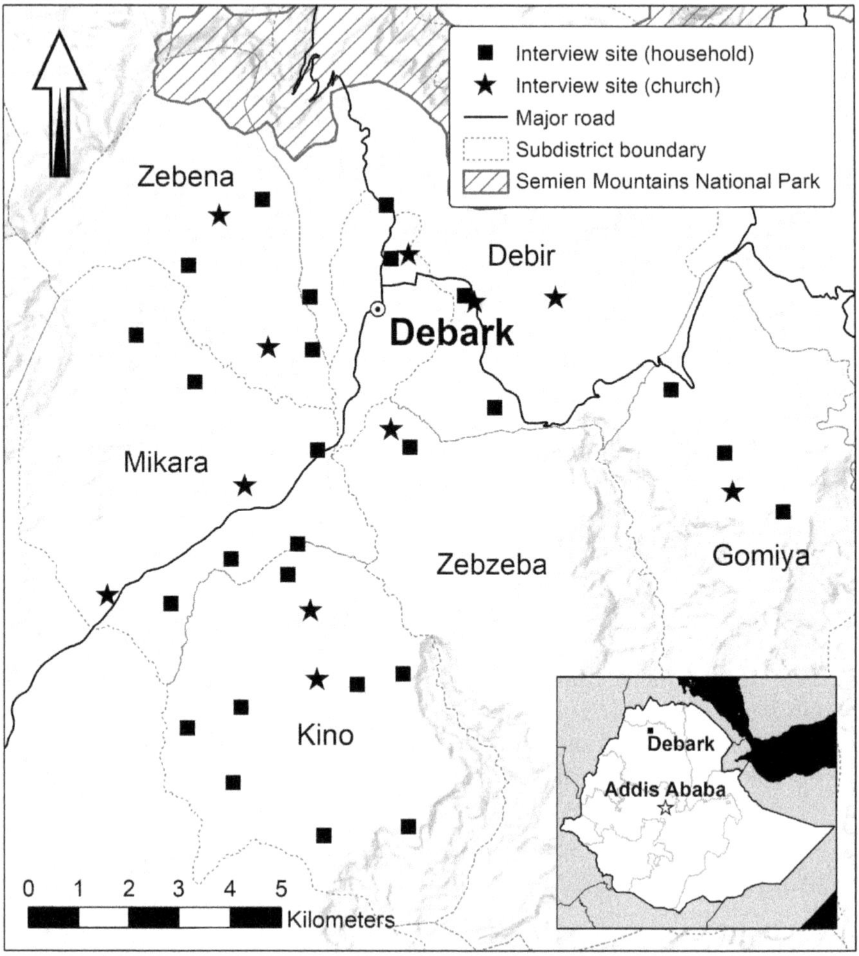

Figure 15.1 Map of research sites in the Debark District of northwestern Ethiopia.

numbers of Protestant Christians and Muslims, particularly in Debark town. Eighty-seven percent of the District's population lives in its rural areas, where they engage in mixed crop and livestock production. The most common field crops are barley (*Hordeum vulgare*), wheat (*Triticum* spp.), flaxseed (*Linum usitatissimum*), faba bean (*Vicia faba*), and field pea (*Pisum sativum*). Farmers also produce a variety of vegetables and root crops in irrigated fields and home gardens. Most farmers also keep domesticated animals, including horses, donkeys, oxen, sheep, goats, chickens, and bees. Most Debark farmers are subsistence-oriented, meaning they produce what they need for their own families and sell any surplus at the market in Debark town.

Data collection

Findings reported here are part of the first author's dissertation research, which focused on farmers' ecological relations with plant diversity and their contributions to the local food system (Ruelle 2015; Ruelle et al. 2018, 2019). Field research was conducted from 2011 to 2014 in six rural subdistricts (*kebeles*) surrounding Debark town, in villages located between 2,800 and 3,000 meters above sea level. Permission to conduct research in these communities was provided by the federal Ethiopian Biodiversity Institute, the Debark District Administration, and District Extension offices. All participants in the research provided informed oral consent prior to each interview or other activity, following protocols approved by the Institutional Review Board at Cornell University.

Altogether, 30 families in 28 villages were interviewed as part of this research. Initial interviews included free listing, an ethnobotanical technique in which participants list all of the plants known to occur within their village and surrounding area, followed by questions about the use and value of each species. All interviews were documented in field notes, which were translated back into Amharic and reviewed with farmers during a subsequent visit to ensure their accuracy.

Participatory mapping was conducted to understand how farmers perceive the abundance and distribution of plants in their landscape. During the first field season of 2011, participatory mapping was facilitated by an extension agent from the Debark District Extension Office, who directed participants to construct a model of their landscape using stones and twigs, which served as the basis for a paper map. During the second field season, we followed a different approach, asking younger members of each family (who had experience drawing pictures at school) to illustrate their home, village, and surrounding landscape. In both cases, farmers were asked to identify the woody plant species depicted on their maps, and names were recorded either directly on the map or elsewhere in field notes.

Direct matrix ranking was conducted with two individuals in each of the six subdistricts. These were individuals who were identified as knowledgeable about woody plant diversity based on their participation in an earlier survey (Tefera et al. 2014). The five most frequently mentioned species in that survey were included in the activity according to eight criteria including: use as construction, fencing, fodder, and fuel wood; significance as a source of income; availability of saplings; growth rate; and multifunctionality.

In the course of our research, we observed that the highest densities of indigenous trees were associated with Ethiopian Orthodox churches. We decided to conduct additional interviews with members of the clergy about the trees around their churches. Permission to conduct interviews at 11 local churches was granted by church administrators in Debark. In addition to listing the trees in their church compound, we asked about the value of trees to the church, including their use in religious practices and symbolic significance. As with farmers, interviews were documented in field notes; however, our letters of permission limited us to one visit to each church, so we have not been able to validate those notes as we had done with farmers.

Finally, a vegetation survey was conducted to estimate the density of eucalyptus and other woody plant species within the study area. Random points were generated in a geographic information system (GIS) (ArcMap 3.0) and transferred to a global positioning system (GPS) to locate and establish 113 circular plots with a 10-meter radius. All woody plants greater than 10 cm in height were identified. Voucher specimens of each species were stored in a portable plant press and transported to the National Herbarium at Addis Ababa University for identification based on the Flora of Ethiopia, comparison with existing herbarium specimens, and consultation with taxonomists in the Department of Plant Biology and Biodiversity Management.

Data analysis

The frequency of free-listing was calculated for each species by dividing the number of households who mentioned that species by the total number interviewed. Field notes from interviews were qualitatively analyzed using open iterative coding in Atlas.ti (versions 7 and 8). Codes included the local names of species, plant uses, and other values mentioned by interviewees. Similarly, maps generated with farmers were subjected to qualitative analysis of salient features, including the number and relative sizes of features and associated woody plant species.

Direct matrix ranking data for five woody species was analyzed with the Plackett-Luce model extended to accommodate partial rankings and ties in the 'PlackettLuce' package (Turner et al. 2020) in R (version 4.03). Models pertaining to each criterion were fitted using the 'iterative scaling' algorithm. Prior strength (the 'npseudo' parameter) was set at the default value of 0.5. Quasi-standard errors were calculated using the 'qvcalc' package (Firth 2020) in R, using eucalyptus as our reference.

Results and discussion

The value of eucalyptus and indigenous trees

To visit rural communities in Debark means walking along roads lined with eucalyptus. If it has rained a few days before, the air is perfumed with their fresh scent. If it is dry, their leaves crackle underfoot. Drawing close to a village, their trunks are woven with eucalyptus branches, forming basket-like wattle fences that obscure yards and houses from view. Standing at an opening in this fence, greeting the family and asking if we might conduct an interview about plants, we see an Elder sitting in the yard, swatting flies with a small bundle of eucalyptus leaves. As we enter, a woman gathers eucalyptus twigs to start a fire to prepare fresh coffee. We sit and talk in the yard, usually in the shade of eucalyptus trees, and see the family's house is constructed of eucalyptus poles, as is the traditional *gojo bet* where the women prepare meals. Eucalyptus is omnipresent and widely used for so many purposes, that we are tempted to describe Debark's farmers as 'people of eucalyptus'.

During our initial interviews, we identified 59 species of trees, shrubs, and lianas. Of these, we consider 13 species (22%) domesticated because all participants told us they are always planted. Eucalyptus belongs to this category. Another 26 species (44%) we determine to be non-domesticated, because participants agree they grow on their own (i.e., are never planted). The remaining 20 species (34%) are sometimes planted and sometimes grow on their own; we categorize these as semi-domesticated. Most of the species in this third category are indigenous species that have been brought into cultivation, but the category includes a few domesticated species from elsewhere that have 'escaped' into uncultivated spaces. This is not the case with eucalyptus, which has never been observed to reproduce on its own, at least by the farmers we interviewed. Furthermore, unlike many trees, eucalyptus can resprout quickly (coppice) from its stump after harvest, so there is rarely a need to replant it.

When asked to name the plants found in their village and surrounding landscape, eucalyptus was the only woody species named by every farmer (see Table 15.1). In fact, a eucalyptus tree was visible from every one of our interview sites. We did not record the order in which plants are listed (some ethnobotanists use this information to evaluate the salience of species in the minds of participants), but in several cases, eucalyptus was not named until we asked if there were other plants that we could see from where we were sitting. It seems eucalyptus is so ubiquitous, its presence so unremarkable, that some farmers had forgotten to mention it.

According to an Elder and respected local historian (Abasqe, personal communication, October 12, 2012), eucalyptus did not arrive in Debark until the 1940s. According to this source, eucalyptus was planted in Gonder during the Italian occupation (1936 to 1941) and later conveyed to Debark. Almost all the eucalyptus trees we encountered during field work were *Eucalyptus globulus*, which some farmers refer to as *talyan bahirzaf* ('Italian eucalyptus'), lending further support to this historical narrative. Most of the voucher specimens we collected were identified as *E. globulus* subsp. *globulus*, one of the first types of eucalyptus brought to Ethiopia, recorded by 1895 (Friis 1995). We also collected specimens of *E. camaldulensis*, one of the more common species of eucalyptus in Ethiopia, but rare in Debark. Farmers who had encountered both species described *E. globulus* as *nech* ('white') and *E. camaldulensis* as *qey* ('red'). To be clear, whenever a farmer referred to *bahirzaf*, we could safely assume he or she was speaking of *E. globulus*.

Aside from eucalyptus, four other trees were mentioned by a majority of farmers, that is, during at least half of our interviews. We will briefly introduce these species because they serve as comparisons to help us understand the value of eucalyptus. First, *Acacia abyssinica*, known in Debark as *girar*, was named during 83% of interviews. It is the only indigenous species of acacia we

Table 15.1 Most frequently listed woody plant species in the Debark District of northwestern Ethiopia (n=30)

Local vernacular Amharic name	*Scientific name*	*Family*	*Cultivation*	*Free-list %*
Bahirzaf	*Eucalyptus globulus*	Myrtaceae	domesticated	100%
Girar (habesha)	*Acacia abyssinica*	Fabaceae	semi-domesticated	83%
Gesho	*Rhamnus prinoides*	Rhamnaceae	domesticated	77%
Tsid (habesha)	*Juniperus procera*	Cupressaceae	semi-domesticated	63%
Weyra	*Olea europaea* subsp. *cuspidata*	Oleaceae	semi-domesticated	50%

encountered in Debark. Some farmers listed other exotic species of acacia; to distinguish it from those, they referred to *A. abyssinica* as *yehabesha* ('Ethiopian') *girar.* While a few farmers said they had planted acacia, it often grows spontaneously from the soil seed bank. Farmers tend to tolerate or encourage seedlings that appear in desirable spots within their crop fields, home gardens, or yards. They will plow around it and may go as far as protecting it with a fence or stone wall. Like other species in its genus, *A. abyssinica* has a broad, umbrella-like shape that provides shade for animals and the people that take care of them, and its pods are considered nutritious for sheep, especially lambs. Acacia is therefore associated with communal grazing areas. For example, a 50-hectare forest known as *atie ketema* ('the Emperor's City') is located two kilometers south of Debark town and comprised almost entirely of *A. abyssinica* trees. Acacia branches are used to construct fences around houses and gardens; its sharp spines deter animals and thieves from entering gardens and houses. Acacia wood was said to make excellent charcoal, although the use of indigenous trees for fuel is officially prohibited.

Gesho (*Rhamnus prinoides*) was mentioned during 77% of interviews and was often observed in farmers' home gardens. Although *gesho* is indigenous, Debark farmers described it as domesticated because it is widely planted and no longer observed growing on its own. *Gesho* leaves and stems are dried, crushed, and ground into a powder that is used as a bittering agent for traditional beer (*tella*). The dried chopped woody branches are also used to prepare *tej* (mead). *Gesho* is sold at the Debark market as bundles of fresh leaves and green stems, as dried crushed leaves, or as a powder.

African pencil cedar (*Juniperus procera*), known locally as *tsid,* was named during 63% of interviews. The name *tsid* is also used for exotic conifers (e.g., *Cupressus lusitanica* and *Casuarina* sp.); farmers who mentioned other conifers referred to *Juniperus procera* as *yehabesha tsid*. *Juniperus procera* is often the dominant species in Afromontane forests (Friis et al. 2010). Cedar is closely associated with Ethiopian Orthodox churches, and the forests surrounding those churches contain the highest densities of them. The priests we interviewed in Debark told us the cedars around their churches had been planted when the church buildings were constructed, as long ago as the 15th century. Cedars therefore bear witness to the history of the church and testify to the durability of Orthodox Christian religious practice and its stewardship of sacred spaces. Farmers said that the cedars in their farmland also belong to the church, so it is forbidden for them to use them. Priests reported that cedars are only ever used to construct and repair church buildings, especially the doors of church buildings and gates to the church compound.

Finally, half of the households we interviewed named African olive (*Olea europaea* subsp. *cuspidata*), known as *weyra*. Farmers reported that olive is sometimes cultivated from seed, but typically grows on its own from the soil seed bank and is tolerated or encouraged when a seedling appears at a desirable location. Olive is an abundant midstory species in Afromontane forests (Friis et al. 2010). When farmers find an olive seedling, they will often construct fences and stone walls to protect it from livestock. Like cedar, olive is closely associated with Orthodox churches, where we observed its greatest abundance. Olive leaves are used by the priests to wash the feet of the faithful on the *tsilot hamus* (Maundy Thursday, four days before Easter). Unlike cedar, olive is frequently used by farmers for non-religious purposes. Its hard wood is used to fashion the tip of a traditional plow (*maresha*). Small green twigs from olive trees are used as toothbrushes and the leaves, young twigs, and splints are used for fumigating household containers. Olive wood is known to make excellent firewood and charcoal, although using it for this purpose has become illegal in recent decades.

Direct matrix ranking was used to compare eucalyptus with acacia, cedar, *gesho*, and an indigenous shrub known as *mogne qitel (Solanecio gigas)* according to eight criteria (Figure 15.2). *Mogne qitel* (meaning 'fool's leaf' in Amharic) is commonly planted along fence lines and produces large leaves used in the process of sprouting barley as traditional malt to make local beer. For seven of

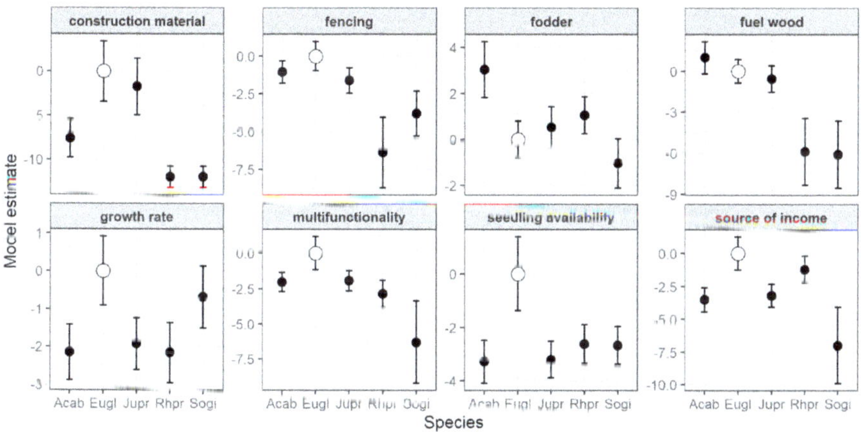

Figure 15.2 Ranking of five woody species in the Debark District of northwestern Ethiopia according to eight criteria (n=12). Acab = *Acacia abyssinica*, Eugl = *Eucalyptus globulus*, Jupr = *Juniperus procera*, Rhpr = *Rhamnus prinoides*, and Sogi = *Solanecio gigas*. Error bars represent two quasi-standard errors.

eight criteria, eucalyptus was among the highest ranked species. First, eucalyptus is known to grow faster than most other trees; farmers ranked its growth rate significantly higher than that of acacia, cedar, and *gesho*. Some farmers reported they are able to harvest eucalyptus poles five or six years after planting. The availability of eucalyptus seedlings was ranked significantly higher than the other four species. Indeed, we observed that eucalyptus seedlings are available for purchase at the Debark market and through the district extension office. Although government nurseries produce seedlings of other species (we toured one nursery west of Debark town), when we purchased seedlings from the Debark extension office, eucalyptus was the only species available.

As a construction material, eucalyptus and cedar were ranked significantly higher than the other three species (acacia is rarely used in construction, and *gesho* and *mogne qitel* are never used for this purpose). Eucalyptus and acacia were ranked highest for use as fences. While the spines on acacia branches serve as a deterrent, the pliability of eucalyptus and the fact that its leaves stay attached after the branches have dried creates a dense wattle fence. Eucalyptus, acacia, and cedar were ranked highest for their use as fuel wood, although eucalyptus is most frequently used, since harvesting indigenous trees is prohibited. We observed that most rural families used eucalyptus as their main fuel source, whereas in Debark town, charcoal is commonly used. The only use for which eucalyptus was ranked significantly lower than other species was as a source of fodder. Its fruits and leaves are known to be unpalatable for livestock, whereas acacia is browsed by goats and its pods are fed to sheep.

It follows that eucalyptus was ranked higher than all other species for its multifunctionality. During interviews, farmers described more uses for eucalyptus than for most other species (see Figure 15.3). Over the course of interviews, we documented 12 distinct uses. In addition to the uses mentioned

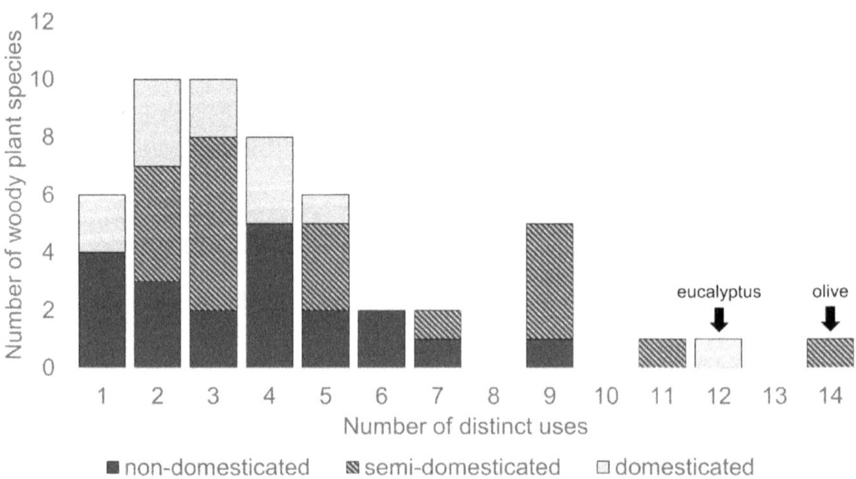

Figure 15.3 Multifunctionality of woody plant species in Debark, Ethiopia.

earlier, farmers reported that eucalyptus blossoms are an important source of nectar for bees; its leaves are used as a medicine to treat nasal congestion; its pliable bark is used as a fastener for fences and walls; the branches are used as broom handles; and the wood is used to fashion the handles of a traditional plow and to construct risers to protect food stores from ants. Eucalyptus is also used in soil and water conservation: the trunks are used to reinforce soil berms that slow the loss of topsoil on steep farmland, and it is planted to retain soil and to prevent waterlogging. Only olive, which was not included in our direct matrix ranking activity, has been identified as having a greater number (14) of uses. We note that some uses are more important than others, so simply counting the number of distinct uses may not represent farmers' own assessment of a species' multifunctionality.

Eucalyptus and *gesho* were ranked significantly higher than the other three species as a source of income. Indeed, these two species were observed for sale at the Debark market throughout the year. A few farmers told us that they were replacing their field crops with eucalyptus because it is a more reliable source of income. Throughout the study area, farmers reported that soil fertility was declining. Eucalyptus appears to thrive on depleted soils, providing an alternative source of income for households who can afford to forego crop production in the interim five to seven years.

Participatory mapping with farming families confirmed that eucalyptus is so ubiquitous that its presence is not remarkable. Eucalyptus was included by farmers in every map, most often around their homestead, along roads, or in woodlots owned by the family. Not surprisingly, cedar and olive were drawn close to churches, *gesho* in home gardens, and acacia in grazing areas. While depictions of indigenous trees were somewhat proportional to their occurrence, farmers underrepresented the abundance of eucalyptus, drawing far fewer eucalyptus trees than we observed in the landscape.

The vegetation survey found that eucalyptus was present in 20% of 113 random plots, far more than any other tree species (see Table 15.1). Acacia had the next highest plot frequency recorded (in 5% of plots), followed by cedar (only 2%). In addition, eucalyptus exhibited the highest density of any species (477 individuals per hectare), more than seven times the density of cedar and 55 times that of acacia. Cedar was found at high densities in a small number of plots, whereas acacia was found at lower densities but in more plots. *Gesho* and olive were never detected in any of our survey plots. As reported by farmers, *gesho* is limited to home gardens (which were never encountered) and olive is still much rarer.

The abundance of eucalyptus in the landscape suggests its value to farmers. Since it is incapable of propagating on its own (except via coppicing), we can assume all eucalyptus trees have been planted. Some plantings, such as those along roadsides and in large woodlots, might have been created by local government initiatives, while those in fields, gardens, and homesteads are likely to have been planted by individual farmers. In other parts of Ethiopia, farmer collectives have played a significant role in planting and tending to eucalyptus woodlots (Dessie et al. 2019). Farmers did not mention those kinds of institutions in Debark,

Table 15.2 Woody plants encountered in a simple random vegetation survey of six sub-districts close to Debark, Ethiopia in 2013, including 113 circular plots of 10-meter radius

Local vernacular Amharic name	*Scientific name*	*Growth form*	*Plot frequency*	*Density (ind/ha)*
Bahir zaf	*Eucalyptus globulus* Labill.	Tree	20.4%	476.6
Embacho	*Rumex nervosus* Vahl	Shrub	11.5%	258.7
Tsid (yehabesha)	*Juniperus procera* Hochst. ex Endl.	Tree	1.8%	63.1
Fyelefej	*Clutia lanceolata* Forssk.	Shrub	1.8%	25.6
Yeferenji tsid	*Cupressus lusitanica* Mill.	Tree	1.8%	15.8
Qebero wetet	*Euphorbia petitiana* A. Rich.	Shrub	3.5%	15.2
Girar (yehabesha)	*Acacia abyssinica* Benth.	Tree	5.3%	8.7
Amija	*Hypericum* sp.	Shrub	3.5%	4.5
Imbuay	*Solanum marginatum* L.f.	Shrub	3.5%	2.3
Atat	*Maytenus obscura* (A. Rich.) Cufod.	Tree	1.8%	1.7
Haregresa	*Zehnaria scabra* Sond.	Climber	1.8%	1.1
Unknown	*Mikaniopsis clematoides* (Sch.Bip. ex A.Rich.) Milne-Redh.	Climber	0.9%	1.1
Unknown	*Vernonia bipontini* Vatke	Shrub	1.8%	0.8
Mogne qitel	*Solanecio gigas* (Vatke) C.Jeffrey	Shrub	0.9%	0.8
Yeferenji girar	*Acacia mearnsii* De Wild.	Tree	0.9%	0.6
Qeret	*Osyris lanceolata* Hochst. & Steud.	Tree	0.9%	0.6
Qaga	*Rosa abyssinica* R.Br.	Shrub	0.9%	0.3
Unknown	*Anthospermum pachyrrhizum* Hiern	Shrub	0.9%	0.3

although we did not ask farmers about them. Furthermore, decisions to plant eucalyptus are likely informed by other factors that we have not captured in this research, for example, pressure or incentives from extension agents and changes in market prices. In any case, farmers' relationships with eucalyptus are more complicated than we might assume based on its abundance in their landscape.

Farmers' concerns about eucalyptus

Many of our conversations revealed ambivalence – that is, the coexistence of opposing feelings – about these trees, even among farmers who had decided to plant eucalyptus on their own land. Despite eucalyptus playing an important role in their lives, including its use in many aspects of their food system, Debark farmers do not consider it part of their culture. First there is the matter of its Amharic name, *bahir zaf*, which translates directly as 'sea tree', but may more accurately be interpreted as 'overseas tree' given its foreign provenence (Pankhurst 1995). The name serves as a constant reminder that eucalyptus is not of this place, as do many Amharic names for plants. More commonly, the names of plants introduced from other parts of the world take the names of similar indigenous species but predicated with the adjective *yeferenji*, meaning 'foreign'. For example, *Cupressus lusitanica* (Mexican cypress) is called *yeferenji tsid* or 'foreign cedar' based on its similarity to *Juniperus procera*. Some introduced species are not labeled as such, and in those cases, their foreign origins appear to be forgotten. For example, Ethiopians consider chili peppers to be an essential part of their cuisine, and their names do not refer to their origins elsewhere (in Central and South America). It may be a matter of time before eucalyptus, like chili peppers, are considered culturally Ethiopian. To explore this argument a little further, Debark farmers use the same word (*behalawi*) for 'culture' and 'tradition', implying that cultural phenomena are established through practice over time. It will be interesting to see how much time will pass before eucalyptus – despite its name – is considered traditional.

The other reasons farmers are skeptical about eucalyptus have to do with its negative impacts on the agroecosystem. Interestingly, while many ecologists have expressed concerns about the allelopathic effects of eucalyptus (i.e., the release of chemicals through leaf litter or root systems that retard the growth of other plants), none of the farmers we interviewed shared their concern. Furthermore, because its branches are relatively short and upright, it does not produce as much shade as trees with spreading habits like acacia, reducing its effects on crop production when used as a border planting. However, farmers throughout our study area observed that eucalyptus uses more water than other trees. Farmers said it was responsible for 'drying out' their landscape, and several pointed to reduced flow in rivers and streams, particularly during the dry season. Farmers' observations that eucalyptus has negative impacts on water resources have been confirmed in other parts of Ethiopia (Chanie et al. 2013). Concerns about water security are exacerbated by the effects of global climate change, which is causing a shorter, later, and less reliable rainy season.

In addition to impacts on water resources, Debark farmers are concerned that eucalyptus is replacing their indigenous tree species. The same Elder local historian mentioned earlier recited a poem that described how eucalyptus had become the 'new king', taking the place of indigenous trees like cedar and olive. On the one hand, the recognition of a eucalyptus as a new king of trees in the eyes of these communities confirm its significance and may even suggest the development of its cultural value. On the other hand, the new king may have usurped the throne from its rightful heirs. Many of the farmers we interviewed shared a view that the proliferation of eucalyptus had negative implications for indigenous trees. Most explained that, given a lack of space, they are faced with choosing between tree species, and decisions to plant eucalyptus are driven by its rapid growth rate and other desirable qualities. However, reliance on eucalyptus makes farmers more vulnerable to pests, diseases, and other climate-related hazards that might one day affect the tree species. Recent studies have suggested innovative forestry systems that mix eucalyptus and indigenous trees to enhance the resilience of the agroecosystem (Amazonas et al. 2018).

A few farmers referred to a government policy forbidding the use of indigenous trees and said this deterred them from planting those species. We spoke with local government officials but were not able to clarify whether farmers were allowed to harvest indigenous trees they had planted themselves. Furthermore, we could not determine if officials kept a record of tree plantings, which would be necessary to determine who could legally use them. Given the lack of clarity about this rule and the difficulties associated with enforcement, farmers hesitate to plant indigenous trees they may not be permitted to use. In some ways, the abundance of eucalyptus in the landscape does not reflect the desires of farmers, but rather the unintended consequences of policies intended to protect indigenous species.

In our conversations about trees, many farmers pointed to the small patches of forests that surround their local church, which serve as refugia for many indigenous species that are difficult or impossible to find in the surrounding landscape. As we have described elsewhere (Ruelle et al. 2018), our conversations with the clergy indicated that many of the indigenous trees within Debark's church forests have been planted, including some of the oldest cedars and olives. Some priests described a policy to maintain the population of indigenous trees within their forest: whenever they harvest a tree, they plant two seedlings in its place. Interestingly, every one of the 11 church forests we visited contained eucalyptus. The clergy and lay people explained that they plant eucalyptus for two reasons: first to raise money for the church, but also to use in place of cedar when they repair church buildings or gates. In other words, planting eucalyptus is intended to reduce pressure on older, indigenous trees of greater cultural significance.

The arrangement of trees within Debark's church forests provides important insights into the value of eucalyptus relative to other trees. Ethiopian Orthodox churches are arranged as a series of concentric circles, both within the main church building and extending to the church compound and outward into the surrounding landscape (Heldman 1992). In a room at the center of the church lies the sacred *tabot* (a recreation of the Arc of the Covenant). Only priests may enter this room in their performance of the Eucharist. Others are allowed to enter the second or third circles of the church, depending on their role in the liturgy and their observance of rituals such as fasting. Many church attendees participate in the service from outside, standing or sitting on benches in the shade of trees encircling the church. At the center of most churches, we find the oldest cedars and olives, whereas eucalyptus is limited to the periphery, the last layer of the church forest. This arrangement has a practical explanation: if the clergy decide to cut one of the eucalyptus trees, they prefer to do so where it will not pose any risk to the church buildings. But there is no denying the symbolic implications of eucalyptus at the edge of the church forest, linking the church to the agricultural economy beyond, but not allowed within the sacred space at the heart of Ethiopian Orthodox practice.

Reconciling folk value

In our interviews in Debark, we did not detect an ideological divide between those who are 'for' or 'against' planting eucalyptus. Rather, most of the farmers we interviewed expressed conflicting feelings about these trees. While this may be a rather extreme case (in which the tree in question is both extremely important and threatening), we would expect other farmers to hold multiple, even irreconcilable attitudes toward the trees in their landscapes. Such ambivalence can inform our understanding of folk value.

We propose that the folk value of a species includes 1) utilitarian values associated with direct use by the community, 2) ecological values based on effects on other organisms and their shared habitat, and 3) symbolic, spiritual, and other intangible cultural values. (In many ways, these categories correspond to the ecosystem services framework introduced in the outset only supporting and regulating services combined as ecological values.) In the case of eucalyptus, the utilitarian values are high, cultural values are low, and ecological values are low or even negative. We may be tempted to unite these three values into some sort of 'folk value index'; however, this would result in its cultural and ecological values offsetting its utilitarian significance, resulting in a measurement similar to that of a species that has low values in all three domains. Therefore, we cannot describe folk value along a theoretical continuum (i.e., as 'high' or 'low'). Rather, folk value must be a multidimensional concept to account for the complexity of human ecological relationships.

Conclusion

Although introduced less than a century ago, eucalyptus is indeed a 'new king' as the dominant component of woody vegetation in Debark. Eucalyptus is not only abundant throughout Debark's agricultural landscape, but it also makes vital contributions to the food system and livelihoods of farming families. Our activities with farmers confirm that it is ranked significantly higher than indigenous trees as a construction material, fuel wood, fencing material, and source of income. Yet, despite its many benefits, farmers maintain an ambivalent attitude toward eucalyptus. Even though it has come to play an important role in their daily lives – from the fire they use to prepare coffee in the morning to the roofs over their heads at night – most farmers do not recognize eucalyptus as part of their culture. This resistance is related to its foreign origins and its negative impacts on their agroecosystem (especially surface water resources). Although it was introduced to ease mounting pressure on native trees, farmers hold it responsible for the decline of those species, including culturally significant cedar and olive. While its inclusion in the sacred spaces of the Ethiopian church might signal a growing acceptance of eucalyptus, its placement at the periphery of the church forest signifies its tentative cultural status. To understand the folk value of eucalyptus is to acknowledge the tension between its material use, agroecological impacts, and cultural significance.

Recommendations

The introduction of eucalyptus and other fast-growing multipurpose trees was intended – at least in part – to reduce demand for indigenous tree products. Indeed, planting eucalyptus in Debark may be helping to protect trees in nearby natural forests, such as those protected within Semien Mountains National

Park. However, according to the farmers we interviewed, the proliferation of eucalyptus is driving a decline of indigenous trees. When an indigenous tree dies, most farmers opt to plant eucalyptus in its place. Such decisions are not surprising, given that eucalyptus is the fastest growing and one of the most multifunctional tree species available. Furthermore, replacement of indigenous trees with eucalyptus is reinforced by 1) low availability of indigenous seedlings as compared to eucalyptus from government nurseries, and 2) policies forbidding cutting of indigenous tree species, thereby disincentivizing planting them.

We envision the Ethiopian Orthodox church as playing an important role in increasing the availability of indigenous seedlings in rural farming communities. Some of the priests we interviewed already know how to collect and germinate seeds from indigenous trees, which they are using to restore and even expand the forests around their churches. Extension agents and civil society organizations could provide additional training for members of the clergy and other church employees to create small nurseries for indigenous trees. Sales of seedlings could raise much needed funds for church repairs and other activities, perhaps reducing the need to plant eucalyptus to sell at market.

Rules and regulations pertaining to the use of indigenous trees need to be clarified and communicated to the public. Many of the farmers we interviewed believe they could be fined for cutting any indigenous tree, even one they planted it themselves. The uncertainty that they will be allowed to use a tree they planted reduces its value relative to eucalyptus, which they know they will be able to use. One way to document who planted which trees (and therefore who is permitted to cut them) is for farmers and extension agents to document locations using the GPS function on a mobile phone. The same technology could be used to encourage farmers to plant indigenous trees and provide guidance as to optimal planting times and locations.

Future research should measure the effects of eucalyptus on water resources and develop strategies to avoid negative impacts. Debark farmers suggested that limiting new eucalyptus planting to hillsides might reduce their impacts on rivers and streams, and may also slow soil erosion, a major concern throughout northern Ethiopia. Engaging farmers in collaborative research could generate new knowledge about the hydrological effects of eucalyptus plantings and initiate a process whereby communities decide how to manage its proliferation.

References

Abasqe. 2012. *Interview with Ato Abasqe on History of Trees in Debark*. Personal communication, October 12.

Adimassu, Z., A. Kessler, C. Yirga, and L. Stroosnijder. 2010. Mismatches between Farmers and Experts on Eucalyptus in Meskan Woreda, Ethiopia. In *Eucalyptus Species Management, History, Status and Trends in Ethiopia. Proceedings from the Congress*, edited by L. Gil, W. Taddese, E. Tolosana, and R. Lopez, pp. 146–159. Ethiopian Institute of Agricultural Research, Addis Ababa, Ethiopia.

Amazonas, N. T., D. I. Forrester, C. C. Silva, D. R. A. Almeida, R. R. Rodrigues, and P. H. S. Brancalion. 2018. High Diversity Mixed Plantations of Eucalyptus and Native Trees: An

Interface between Production and Restoration for the Tropics. *Forest Ecology and Management* 417: 247–256. https://doi.org/10.1016/j.foreco.2018.03.015.

Central Statistics Agency. (2008). *Summary and Statistical Report of the 2007 Population and Housing Census*. Population Census Commission of the Federal Democratic Republic of Ethiopia.

Chanie, T., A. S. Collick, E. Adgo, C. J. Lehmann, and T. S. Steenhuis. 2013. Eco-hydrological Impacts of Eucalyptus in the Semi Humid Ethiopian Highlands: The Lake Tana Plain. *Journal of Hydrology and Hydromechanics* 61(1): 21–29b. https://doi.org/10.2478/johh-2013-0004.

Chekole, G., Z. Asfaw, and E. Kelbessa. 2015. Ethnobotanical Study of Medicinal Plants in the Environs of Tara-gedam and Amba Remnant Forests of Libo Kemkem District, Northwest Ethiopia. *Journal of Ethnobiology and Ethnomedicine* 11(1): 4. https://doi.org/10.1186/1746-4269-11-4.

Demissew, S., I. Friis, and O. Weber. 2021. Diversity and Endemism of the Flora of Ethiopia and Eritrea: State of Knowledge and Future Perspectives. *Rendiconti Lincei. Scienze Fisiche e Naturali*. https://doi.org/10.1007/s12210-021-01027-8.

Dessie, A. B., A. A. Abtew, and A. D. Koye. 2019. Analysis of Smallholder Farmers' Cooperation in Eucalyptus Woodlot Production in Wegera District, Northern Ethiopia. *Small-Scale Forestry* 18(3): 291–308. https://doi.org/10.1007/s11842-019-09418-4.

Dessie, G., and T. Erkossa. 2011. *Eucalyptus in East Africa: Socio-Economic and Environmental Issues*. Working Paper FP46/E; Forestry Department, p. 30. Food and Agriculture Organization of the United Nations.

Feyisa, D., E. Kissi, and Z. Kebebew. 2018. Rethinking Eucalyptus globulus Labill. Based Land Use Systems in Smallholder Farmers Livelihoods: A Case of Kolobo Watershed, West Shewa, Ethiopia. *Ekológia (Bratislava)* 37(1): 57–68. https://doi.org/10.2478/eko-2018-0006.

Firth, D. 2020. *qvcalc: Quasi Variances for Factor Effects in Statistical Models* (R package version 1.0.2) [Computer software]. Available at https://CRAN.R-project.org/package=qvcalc

Franzel, S. C., H. Jaenicke, and W. G. Janssen. 1996. *Choosing the Right Trees: Setting Priorities for Multipurpose Tree Improvement*. International Service for National Agricultural Research, The Hague, Netherlands.

Friis, I. 1995. Myrtaceae. In *Flora of Ethiopia and Eritrea: Volume 2 Part 2 (Canellaceae to Euphorbiaceae): Volume 2, Part 2*, edited by S. Edwards, M. Tadesse, and H. Inga, pp. 71–106. The National Herbarium, Addis Ababa, Ethiopia and Uppsala, Sweden.

Friis, I., Sebsebe Demissew, and P. Von Breugel. 2010. *Atlas of the Potential Vegetation of Ethiopia*. Addis Ababa University Press and Shama Books, Addis Ababa, Ethiopia.

Heldman, M. E. 1992. Architectural Symbolism, Sacred Geography and the Ethiopian Church. *Journal of Religion in Africa* 22(3): 222. https://doi.org/10.2307/1580918.

Jagger, P., and J. Pender. 2003. The Role of Trees for Sustainable Management of Less-favored Lands: The Case of Eucalyptus in Ethiopia. *Forest Policy and Economics* 5(1): 83–95. https://doi.org/10.1016/S1389-9341(01)00078-8.

Kelbessa, E., and S. Demissew. 2014. Diversity of Vascular Plant Taxa of the Flora of Ethiopia and Eritrea. *Ethiopian Journal of Biological Sciences* 13(Supp.): 37–45.

McCann, J. C. 1997. The Plow and the Forest: Narratives of Deforestation in Ethiopia, 1840–1992. *Environmental History* 2(2): 138. https://doi.org/10.2307/3985505.

Meire, E., A. Frankl, A. De Wulf, M. Haile, J. Deckers, and J. Nyssen. 2013. Land Use and Cover Dynamics in Africa since the Nineteenth Century: Warped Terrestrial Photographs of North Ethiopia. *Regional Environmental Change* 13(3): 717–737. https://doi.org/10.1007/s10113-012-0347-9.

Mekonnen, Z. 2010. Community Opinion, Marketing and Current Debates on Eucalyptus in Huruta District, Arsi Zone of Oromia Region, Ethiopia. In *Eucalyptus Species Management, History, Status and Trends in Ethiopia. Proceedings from the Congress*, edited by L. Gil, W. Tadesse, E. Tolsana, and R. Lopez, pp. 132–145, Ethiopian Institute of Agricultural Research, Addis Ababa, Ethiopia.

Mekoya, A., S. J. Oosting, S. Fernandez-Rivera, and A. J. Van der Zijpp. 2008. Farmers' Perceptions about Exotic Multipurpose Fodder Trees and Constraints to their Adoption. *Agroforestry Systems* 73(2): 141–153. https://doi.org/10.1007/s10457-007-9102-5.

Nath, C. D., G. Schroth, and D. F. R. P. Burslem. 2016. Why Do Farmers Plant More Exotic than Native Trees? A Case Study from the Western Ghats, India. *Agriculture, Ecosystems & Environment* 230: 315–328. https://doi.org/10.1016/j.agee.2016.05.013.

Nyssen, J., M. Haile, J. Naudts, N. Munro, J. Poesen, J. Moeyersons, A. Frankl, J. Deckers, and R. Pankhurst. 2009. Desertification? Northern Ethiopia re-photographed after 140 years. *Science of The Total Environment* 407(8): 2749–2755. https://doi.org/10.1016/j.scitotenv.2008.12.016.

Pankhurst, R. 1995. The History of Deforestation and Afforestation in Ethiopia Prior to World War I. *Northeast African Studies, New Series* 2(1), 119–133.

Reed, J., J. van Vianen, S. Foli, J. Clendenning, K. Yang, M. MacDonald, G. Petrokofsky, C. Padoch, and T. Sunderland. 2017. Trees for Life: The Ecosystem Service Contribution of Trees to Food Production and Livelihoods in the Tropics. *Forest Policy and Economics* 84: 62–71. https://doi.org/10.1016/j.forpol.2017.01.012.

Ruelle, M. L. 2015. *Human-Plant Ecology of an Afromontane Agricultural Landscape: Diversity, Knowledge, and Food Sovereignty in Debark, Ethiopia*. Dissertation, Cornell University. Available at https://hdl.handle.net/1813/40982

Ruelle, M. L., K.-A. Kassam, and Asfaw, Z. (2018). Human Ecology of Sacred Space: Church Forests in the Highlands of Northwestern Ethiopia. *Environmental Conservation* 45(3): 291–300. https://doi.org/10.1017/S0376892917000534.

Ruelle, M. L., K.-A. Kassam, S. J. Morreale, Z. Asfaw, A. G. Power, and T. J. Fahey. 2019. Biocultural Diversity and Food Sovereignty: A Case Study of Human-Plant Relations in Northwestern Ethiopia. *Food Security* 11: 183–199. https://doi.org/10.1007/s12571-019-00888-0.

Tefera, B., M. L. Ruelle, Z. Asfaw, and B. Abraha Tsegay. 2014. Woody Plant Diversity in an Afromontane Agricultural Landscape (Debark District, Northern Ethiopia). *Forests, Trees and Livelihoods* 23(4): 261–279. https://doi.org/10.1080/14728028.2014.942709.

Turner, H. L., J. van Etten, D. Firth, and I. Kosmidis. 2020. Modelling Rankings in R: The PlackettLuce Package. *Computational Statistics* 35(3): 1027–1057. https://doi.org/10.1007/s00180-020-00959-3.

Wøien, H. 1995. Deforestation, Information and Citations: A Comment on Environmental Degradation in Highland Ethiopia. *GeoJournal* 37(4): 501–511.

WorldClim. 2021. *Historical Data*. Available at worldclim.org

Zomer, R. J., A. Trabucco, R. Coe, F. Place, M. van Noordwijk, and J. C. Xu. 2014. *Trees on Farms: An Update and Reanalysis of Agroforestry's Global Extent and Socio-Ecological Characteristics*. World Agroforestry Centre (ICRAF), Nairobi, Kenya. Available at https://doi.org/10.5716/WP14064.PDF

Index

Page numbers in italics indicate a figure and page numbers in bold indicate a table on the corresponding page. Page numbers followed by 'n' indicate a note.

For Product Safety Concerns and Information please contact our EU
representative GPSR@taylorandfrancis.com
Taylor & Francis Verlag GmbH, Kaufingerstraße 24, 80331 München, Germany

www.ingramcontent.com/pod-product-compliance
Ingram Content Group UK Ltd.
Pitfield, Milton Keynes, MK11 3LW, UK
UKHW022316250926
475702UK00046B/884
* 9 7 8 0 3 6 7 3 3 6 1 5 8 *